Advances in Fossil Energy R & D 2019

Topical Conference at the 2019 AIChE Annual Meeting

Orlando, Florida, USA
10 - 15 November 2019

ISBN: 978-1-7138-0562-5

Printed from e-media with permission by:

Curran Associates, Inc.
57 Morehouse Lane
Red Hook, NY 12571

CURRAN ASSOCIATES INC.
proceedings
.com

Some format issues inherent in the e-media version may also appear in this print version.

Printed with permission by Curran Associates, Inc. (2020)

For permission requests, please contact AIChE
at the address below.

AIChE
120 Wall Street, FL 23
New York, NY 10005-4020

Phone: (800) 242-4363
Fax: (203) 775-5177

www.aiche.org

Additional copies of this publication are available from:

Curran Associates, Inc.
57 Morehouse Lane
Red Hook, NY 12571 USA
Phone: 845-758-0400
Fax: 845-758-2633
Email: curran@proceedings.com
Web: www.proceedings.com

Advances in Fossil Energy R & D 2019

Topical Conference at the 2019 AIChE Annual Meeting

Orlando, Florida, USA
10 - 15 November 2019

AIChE
The Global Home of Chemical Engineers

TABLE OF CONTENTS

579741 Process Optimization Using Perturbation Expansion Solutions for Membrane Gas Separation Modules

Simulation and optimization of hybrid membrane gas separation processes, such as those proposed for carbon capture, require high fidelity module performance simulations. Unfortunately, these simulations require iterative solution techniques which can be sensitive to the initial solution guess and thus require extensive iteration (i.e., computational time) or fail to converge.

Perturbation solutions to the isothermal, constant transport property, multi-component cross-flow and counter-current flow performance equations are presented. Predictions from the direct, non-iterative solution are compared to solutions obtained using numerical approximations. Excellent agreement is found in all cases even for systems with broad ranges of recovery for feed gas components. The results also are excellent initial guesses for simulations that account for module non-idealities.

The use of these solutions in optimizing staged membrane systems is discussed. The perturbation solution can dramatically speed up optimization and help identify unfeasible configurations.

558394 Novel Microwave-Enhanced Non-Aqueous Slurry Process for Post-Combustion CO_2 Capture

Sorbent regeneration plays a significant role in the adsorption/desorption cycles needed for effective CO_2 capture. This is because selective and efficient desorption can significantly increase the overall efficiency by allowing quicker sorbent bed turn-over and leading to a sorbent working capacity that approaches the maximum CO_2 capture capacity. Current state of the art methods involves a pressure/thermal swing, which often requires superheated steam for reactive regeneration. This leads to reduced efficiency due to longer regeneration times, incomplete CO_2 desorption and working capacities that are significantly lower than the sorbents maximum uptake capacity. Therefore, traditional absorptive carbon capture process leads to slower and inefficient capture/release stages that also increases cost.

In this work, Basic Immobilized Amine Sorbents (BIAS) were mixed with silicone oil as a novel slurry. These mixtures have been developed for post-combustion CO_2 capture to promote absorption/desorption process. This hydrophobic physical slurry has demonstrated rapid regeneration kinetics through the application of microwave (MW) technology. Results of infrared images showed that silicone oil barely absorbed the microwave energy, while BIAS sorbents, especially CO_2 saturated BIAS sorbents containing polar amine groups, strongly absorbed microwave energy. Experimental results showed that local MW heating can significantly expedite CO_2 release from the sorbent and reduce energy consumption during the regeneration process. FTIR spectra have proven that microwave heating didn't damage BIAS sorbent, because silicone oil can disperse heat and therefore prevent sorbent from deactivation due to overheating. The regeneration processes under different microwave conditions have been systematically investigated.

This slurry process has also expressed improved hydrophobicity, and low or negligible corrosivity. Therefore, this novel MW-enhanced slurry method could provide a fast and highly energy efficient pathway for carbon capture.

567897 Low CO_2 Concentration Post-Combustion Capture: Process Insights

For post-combustion CO_2 technologies applicable to power-plants, amine-based solvent CO_2 capture has seen considerable development for commercial deployment among other available options. Although amine-based solvent CO_2 capture has been in use for several decades in gas sweetening applications for example, a key technological challenge for post-combustion CO_2 capture in fossil-fuel power generating plants is the low CO_2 concentration in the flue gas. Typical concentrations range from 12-14 vol% for coal-fired plants and 3-5 vol% for natural-gas fired plants. The low concentration reduces the driving force for CO_2 transfer from the gas phase into the solvent with its consequent impact on the energy of regeneration of the solvent. With technology suited to both natural and coal-fired gas plants, the even lower drying force for CO_2 transfer from natural gas generated-flue gas would require stringent process conditions to overcome the effect of this limitation such as the potentially higher energy implications for solvent regeneration, with the associated high cost of electricity.

In this work, results from experiments conducted in a 0.1 MWth pilot post-combustion CO_2 capture unit at low CO_2 concentration (~5 vol%) mimicking natural gas-fired plant conditions is presented. Impacts of different liquid circulation, gas flow rate and stripper pressure were examined. The results of the study was compared with findings from studies done with coal-derived flue gas. It is shown that appropriate hydrodynamic conditions in the column are required for the low CO_2 concentration based on reduced liquid circulation rates to enhance solvent performance as a result of the low driving force for capture. The corresponding higher oxygen content in the low-content CO_2 gas did not contribute to increased oxidative solvent degradation compared to the impact of the stripping temperature. Leveraging extensive work done with high CO_2 flue gas conditions (applicable to coal-fired plants), comparisons will highlight applicability of knowledge to natural-gas fired plants together with recommended areas for process enhancement, particularly as these plants increase in prominence for power generation.

576441 Process Modeling and Optimal Synthesis of a Fixed Bed System for CO_2 Capture Using a Diamine-Appended Metal–Organic Framework

Functionalized sorbents offer a potential alternative technology for CO_2 capture. Viable sorbents should have lower regeneration energy, faster kinetics, higher CO_2 selectivity, and higher working capacity for as high as 90% CO_2 capture from the post-combustion flue gas. A novel class of diamine-appended metal–organic frameworks (MOFs) offers a number of these features[1,2,3]. These MOFs exhibit step-shaped isotherms under low partial pressures of CO_2 and lead to higher working capacities under similar temperature and pressure intervals when compared to traditional sorbents. The MOF studied in this work, Mg_2(dobpdc) (dobpdc^{4-} = 4,4′-dioxidobiphenyl-3,3′-dicarboxylate), contains one-dimensional hexagonal channels with unsaturated Mg^{2+} sites and is functionalized by binding the diamine 2,2-dimethyl-1,3-diaminopropane (dmpn) to these sites which then chemisorb CO_2 in cooperative mechanism to form carbamate and carbamic acid species. Along with high working capacities, this functionalized MOF also shows excellent long-term stability and maintains performance under humid conditions, both desirable attributes for flue gas CO_2 capture.[1]

For minimizing the penalty due to CO_2 capture, contactors for these functionalized MOFs should be optimally designed. The step-shaped isotherms exhibited by these MOFs cannot be adequately represented by traditional isotherms due to very complex chemical reaction pathways in these MOFs. A new isotherm model that captures the unique mechanism of CO_2 uptake is developed in this work. Data from thermogravimetric analysis (TGA) are used to develop the kinetic model. A model of the fixed bed experimental set up was developed and used to validate the kinetic model by using the experimental data. The model is then scaled up for CO_2 capture from a 550 MWe net coal-fired power plant. The dynamic, pressure driven, fixed bed contactor model is a 1-d first-principles model that includes mass, momentum and energy conservation equations and accounts for internal and external mass transfer resistances, reaction kinetics, heat transfer between the gas and solid phases, and heat transfer between the gas/solid and an embedded heat exchanger within the reactor. The model was then used to simulate an entire temperature swing adsorption (TSA) cycle for adsorption/desorption of CO_2 from a flue gas source. Due to the step-shaped isotherms and reasonably high heat of reaction for this studied system, efficient heat removal/addition is critical. Therefore optimal design and operation of the embedded heat exchangers by considering the capital and operating costs of the process is critical. However the fixed bed process is a dynamic process due to the cyclic nature of the adsorption/desorption cycles. Therefore a dynamic optimization of the fixed process is performed yielding not only the optimal design of the fixed bed capture process but also the optimal operating conditions.

References:

[1] – Milner, P.J., Siegelman, R.L., Forse, A.C., Gonzalez, M.I., Runcevski, T., Martell, J.D., Reimer, J.A., Long, J.R. A Diaminopropane-Appended Metal-Organic Framework Enabling Efficient CO_2 Capture from Coal Flue Gas via a Mixed Adsorption Mechanism. *Journal of the American Chemical Society*. 2017; 139 (38), 13541-13553

[2] Mcdonald, T.M., et al., Cooperative insertion of CO_2 in diamine-appended metal-organic frameworks. *Nature*. 2015; 519, 303-308

[3] Forse, A.C., Milner, P.J., Lee, J., Redfearn, H.N., Oktawiec, J., Siegelman, R.L., Martell, J.D., Dinakar, B., Porter-Zasada, L.B., Gonzalez, M.I., Neaton, J.B., Long, J.R., Reimer, J.A., Elucidating CO_2 Chemisorption in Diamine-Appended Metal-Organic Frameworks. *J. Am. Chem. Soc*. 2018; 140(51),

18016-18031

579375 Highly Polar Polymers with Superior Membrane CO_2/N_2 Separation Properties for Carbon Capture

Membrane technology has emerged as a potentially economically viable alternative for CO_2 captures from fossil fuel-fired powers, enabled by advanced membrane materials with high CO_2 permeability and high CO_2/N_2 selectivity. Current leading membrane materials usually contain poly(ethylene oxide) (PEO) because the ether oxygen in PEO interacts favorably with CO_2, resulting in high CO_2/N_2 selectivity. Herein we prepare a series of highly branched amorphous polymers containing poly(1,3-dioxolane), which has an O:C ratio of 0.67, higher than 0.5 in PEO. The length of the poly(1,3-dioxolane)-based branches are tuned to yield amorphous nature, and mobile ethoxyl chain end groups are introduced to provide high free volume and gas diffusivity. These ether oxygen-rich polymers exhibit more superior CO_2/N_2 separation properties than the PEO-based materials at practical conditions for flue gas processing, and above the Robeson's upper bound. This work demonstrates that harnessing the interactions between polymers and CO_2 may provide unprecedented opportunities in designing gas separation membranes with robust performance under practical conditions.

560702 Systematic Atomistic Simulations of CO_2 and N_2 Permeabilities in Polymers

Recently, we have developed a computational approach [1,2] to screen the properties of physical solvents for specific gas separation applications through the integration of the NIST database, in-house computational database, in-house Monte Carlo and molecular dynamics simulations, and machine learning models. It is interesting but challenging to extend this computational approach to screen polymers for gas separation applications. The challenges include building a reliable database and identifying and computing feature variables for machine learning of gas permeability.

In this work, we aim to address one of the above challenges, that is, to study the factors (feature variables) which affect CO_2 and N_2 permeability and selectivity from atomistic simulations. We systematically calculated CO_2 and N_2 permeability in several different polymers, such as amorphous poly(ethylene) (PE), polydimethylsiloxane (PDMS), and poly[1-(trimethylsilyl)-1-propyne] (PTMSP), which cover a wide range of CO_2 permeabilities from 10-100,000 Barrer. Gas solubility was calculated both from the modified Widom-insertion method [3] and the advanced continuous fractional component (CFC) [4, 5] osmotic Monte Carlo (MC) simulation. The two methods were found to give consistent CO_2 solubility in PE. Gas diffusivity was obtained from NVE MD simulations using the Einstein relation. The simulated CO_2 solubility, diffusivity, and permeability in PE and PDMS polymers are comparable to the corresponding experimental data [6-9] with similar polymer molecular weight. Our simulation results clearly show that CO_2 diffusivity determines the order of magnitude for CO_2 permeability, while, CO_2/N_2 solubility selectivity determines the gas pair permeability selectivity.

For the CH_3- $(C_2H_4)_n$-CH_3 homologous compounds, it was found that the compound density, CO_2 solubility, diffusivity, and permeability correlate very well with $1/n$ for n between 4-100. This allows us to obtain the values for the PE polymer with n between 700-1400 through extrapolation. When n increases, the homologous compound density increases, while the compound free volume fraction, CO_2 solubility, diffusivity, and permeability decrease. Finally, it was found CO_2 absorption in PE at 20 bar expands the polymer volume by 4%, which in turn increases CO_2 diffusivity and permeability by two-fold. A similar increase in CO_2 permeability at elevated CO_2 pressure in PE was also experimentally determined in the literature [7]. This study will also be extended to other polymers and other gases very easily.

References:

1. Wei Shi, Robert L. Thompson, Megan K. Macala, Kevin Resnik, Janice A. Steckel, Nicholas S. Siefert, and David P. Hopkinson, "**Molecular Simulations of CO_2 and H_2 Solubility, CO_2 Diffusivity, and Solvent Viscosity at 298 K for 27 Commercially Available Physical Solvents**", Invited paper to the *J. Chem. Eng. Data*, DOI: 10.1021/acs.jced.8b01228, 2019
2. Wei Shi, Robert L. Thompson, Megan K. Macala, S. Tiwari, Kevin Resnik, Janice A. Steckel, Nicholas Siefert, David P. Hopkinson, "**Integration of Data Mining, Molecular Modeling, and Machine Learning to Screen Physical Solvents for Gas Separation**", 2018 AIChE Annual meeting, David L. Lawrence Convention Center, Pittsburgh, PA, Oct. 28-Nov.2, 2018.
3. Wei Shi, R. L. Thompson, E. Albenze, J. A. Steckel, H. B. Nulwala, D. R. Luebke. "**Contribution of the acetate anion to CO2 solubility in ionic liquids: theoretical method development and experimental study**", *J. Phys. Chem. B* **2014**, 118, 7388-7394.
4. Wei Shi, E. J. Maginn, "Continuous Fractional Component Monte Carlo: An Adaptive Biasing Method for Open System Atomistic Simulations", *J. Chem. Theory Comput.* **2007**, *3*, 1451-1463.
5. Wei Shi, E. J. Maginn, "**Atomistic Simulation of the Absorption of Carbon Dioxide and Water in the Ionic Liquid 1-n-Hexyl-3-methylimidazolium Bis (Trifluoromethylsulfonyl) imide ([hmim][Tf_2N])**", *J. Phys. Chem. B* **2008**, *112*, 2045-2055.

6. A. S. Michaels, R. B. Parker, Jr. "**Sorption and Flow of Gases in Polyethylene**", *J. Polym. Sci.* **1959**, *XLI*, 53-71.
7. S. A. Stern, S. S. Kulkarni, "**TEST of a "Free-Volume" Model for Gas Permeation through Polymer Membranes. I. Pure CO_2, CH_4, C_2H_4, and C_3H_8 in Polyethylene**", *J. Polym. Sci.* **1983**, *21*, 467-481.
8. H. Lin, B. D. Freeman, "**Gas solubility, diffusivity and permeability in poly(ethylene oxide)**", *J. Membr. Sci.* **2004**, *239*, 105-117.
9. T. C. Merkel, V. I. Bondar, K. Nagai, B. D. Freeman, I. Pinnau, "**Gas Sorption, Diffusion, and Permeation in Poly(dimethylsiloxane)**", *J. Polym. Sci. B: Polym. Phys.* **2000**, *38*, 415–434.

572296 Preparation of FAU Zeolite Membranes on Polydopamine Modified Al_2O_3 Tubes for Dewatering of Amine-Based Post-Combustion CO_2 Capture Solutions

Amine scrubbing is currently regarded as among the most economical technology for post-combustion CO_2 capture. In this process, aqueous amine solvent is circulated between an absorber column for CO_2 absorption and a stripping column for solvent regeneration. To reduce the energy required in the stripping column per mole of CO_2 captured, a selective ceramic membrane is proposed here to dewater and enrich the carbon loading in the amine solution prior to the stripper [1]. In this work, FAU zeolite membranes installed after the lean-rich heat exchanger were investigated as part of a dewatering unit for implementation into the carbon capture process. The concentrated CO_2-rich solution is sent to the top of stripper while the water-rich permeate is joined with the lean amine solvent exiting the bottom of the stripper and is sent back to the absorber. Based on polydopamine (PDA) modified Al_2O_3 substrates, a simple and facile synthesis strategy was developed to prepare highly reproducible FAU zeolite membranes. In this approach, polydopamine was deposited on Al_2O_3 tubes (coated with a 30-μm thick α-alumina layer with an average pore size of 0.05 μm) under varying time periods between 1-20 h for the fabrication of cross-linked polymer layer. The optimized modification time was 4 h to achieve uniform dopamine polymerization on the Al_2O_3 surface. Attributing to its bio-adhesive ability through non-covalent self-assembly and covalent polymerization [2], PDA functioned as a "molecular seed" for the attachment of zeolite onto the Al_2O_3 substrate surface for the formation of a dense and selective membrane layer with a thickness of approximately 3 μm. The physicochemical properties of these zeolite membranes were characterized to understand the hydrothermal synthesis temperature and time on membrane formation by techniques including X-ray diffractometer (XRD), scanning electron microscopy (SEM), and energy dispersive X-ray spectroscopy (EDX). The effects of dewatering test operation temperature (353-413 K) and pressure (0-150 psi) on the membrane dewatering performance were comprehensively investigated in a 5 M monoethanolamine (MEA) solution to work towards an optimized condition. Membranes synthesized at 85 °C for 11 h exhibited fluxes as high as 6.5 kg m^{-2} h^{-1} with rejection rates of 95% at 403 K and 75 psi. In addition, FAU zeolite membrane batches were prepared with a high reproducibility of >90%.

References

[1] X. Li, J. E. Remias, J. K. Neathery, K. Liu, *J. Membr. Sci.*, 2011 (366), 220-228.

[2] C. Zhou, C. Yuan, Y. Zhu, J. Caro, A. Huang, *J. Membr. Sci.*, 2015 (494), 174-181.

588761 Fabrication of Thin Film Composite Membranes using Microporous Polymer Blends and Polybenzimidazole Nanoporous Supports for CO_2/N_2 Separation

Fabrication of Thin Film Composite Membranes using Microporous Polymer Blends and Polybenzimidazole Nanoporous Supports for CO_2/N_2 Separation

Lingxiang Zhu[†,‡,*], Ali K. Sekizkardes[†,‡], Victor A. Kusuma[†,‡], Surendar R. Venna[†,‡]

Shouliang Yi[†,‡], David P. Hopkinson[†,*]

† U.S. Department of Energy, National Energy Technology Laboratory, 626 Cochrans Mill Road, Pittsburgh, Pennsylvania 15236, United States

‡ Leidos Research Support Team, 626 Cochrans Mill Road, Pittsburgh, Pennsylvania 15236, United States

*Correspondence to: D. Hopkinson david.hopkinson@netl.doe.gov, L. Zhu lingxiang.zhu@netl.doe.gov

Industrial gas separations usually use large-scale (~10,000 m^2 or more/plant) thin film composite (TFC) membranes that are comprised of a thin (100 nm - 1 μm) selective layer and a thick (> 20 μm) porous support layer. Fabrication and scale-up of the TFC membranes depend on selective layer material with good thin film forming ability and a porous membrane support with optimized pore structure (*i.e.*, small pore size and high porosity) and thus minimal gas transport resistance. In this study, we fabricated TFC membranes based on a cost-effective polymer of intrinsic microporosity/ether side chain polyphosphazene (PIM-1/MEEP) blend that is a high-performance CO_2-selective membrane material developed in-house recently [1]. PIM-1 and MEEP are soluble and miscible in chloroform, rendering PIM-1/MEEP blend exceptional coatability. To accommodate the thin film coating of PIM-1/MEEP, chloroform-resistant polybenzimidazole (PBI) nanoporous supports were fabricated via a phase-inversion method. These PBI nanoporous supports were characterized by pore size of < 20 nm, surface porosity of *ca.* 15%, and CO_2 permeance of 250,000 GPU at 25 °C. Defect-free sub-micron PIM-1 based TFC membranes were prepared on the PBI supports via a dip coating technique, and their pure- and mixed-gas CO_2/N_2 separation properties were systematically investigated. For example, 1-μm-thick PIM-1/MEEP TFC membranes showed pure-gas CO_2 permeance of *ca.* 800 GPU and CO_2/N_2 selectivity of 28 at 25 °C. Their gas separation performance, membrane stability, and aging behavior in flue gas stream at National Carbon Capture Center would be presented as well.

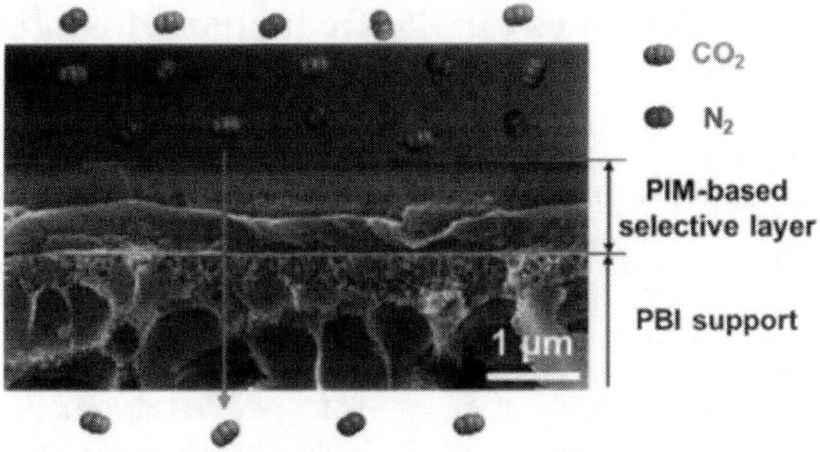

Reference: [1] A. Sekizkardes, V. Kusuma, J. McNally, D. Gidley, K. Resnik, S. Venna, D. Hopkinson, *J. Mater. Chem. A*, 2018,6, 22472-22477

574249 Carbonate Ceramics for the Reduction of Global Carbon Footprint

Utilization of captured carbon dioxide (CO_2) by converting it into valuable products, such as fuels, chemicals, plastics, and building materials is necessary for sufficient carbon management. This presentation will describe a breakthrough process invented and patented by Rutgers University called gas-assisted reactive hydrothermal liquid phase densification (g-rHLPD). When CO_2 is incorporated into the hydrothermal liquid phase to react with synthetic and mineral silicates at temperatures below 100°C, carbonate-bound ceramics with outstanding properties can be produced. These ceramics have applications in cement and advanced composites and are green in terms of their carbon and energetic footprint. This presentation will discuss the utilization of carbonated-$CaSiO_3$ as an ideal binder for a myriad of composite structures and provide a perspective on how its contribution to sustainability can impact a broad spectrum of business in performance enhancing products while significantly reducing the global carbon footprint.

570372 CO_2 Mineralization Kinetics of Portlandite-Enriched Cementing Systems Exposed to Flue Gases

When in contact with flue gases, cementitious binders featuring portlandite ($Ca(OH)_2$) can take up CO_2 by forming stable carbonate minerals (i.e., CO_2 mineralization/carbonation reactions). As these reactions induce cohesive strength between particles, they may be exploited to capture/utilize CO_2 from dilute streams for the fabrication of concrete products with reduced carbon intensities compared to their conventional counterparts. To inform process design, the influences of relative humidity, temperature, and CO_2 concentration on the carbonation kinetics of portlandite within these components must be clearly understood. However, such understanding has been limited by the lack of portlandite kinetics data within conditions relevant to direct flue gas exposure, and due to the complex linkages between the carbonation kinetics of portlandite-enriched components and of the reactants from which they are formed. For instance, CO_2 transport limitations increasingly influence carbonation kinetics as component size increases and are highly dependent on properties such as the porosity and degree of pore saturation with liquid water. Here, we discuss the influences of processing conditions on carbonation kinetics of portlandite-enriched binders at both the particulate scale and the component scale. We assess the sensitivity of the carbonation kinetics of portlandite particulates and portlandite-enriched monoliths to relative humidity, temperature, and CO_2 concentration. The relationships between the CO_2 uptake of portlandite particulates and portlandite-enriched monoliths in relation to their thickness and geometric surface area/volume ratio (SA/V) are elucidated. The temperature rise due to the exothermic carbonation reactions as a function of the component size is also investigated. The findings of this study inform the design of CO_2 mineralization processes in relation to component geometry, aiming towards the scalable production of low-carbon concrete components by direct flue gas utilization.

564092 Implementation of Ion Exchange Processes on Industrial Waste Streams for CO_2 Mineralization

Economic development and population growth will lead to an increase in global energy consumption. Large coal-fired power plants (>500 MW) account for almost half of the total CO_2 emissions from fossil fuel combustion, and nearly 26% of the global fossil fuel and industry CO_2 emissions. Long-term management of the more than 36 Gt of CO_2 that is emitted globally per year is urgently needed. CO_2 mineralization is an attractive method of sequestering CO_2 because it captures CO_2 as an insoluble carbonate mineral, mimicking the natural process of rock weathering. Here, we present a pathway for accelerated CO_2 mineralization enabled by divalent cations sourced from liquid streams.

In this work, we investigate a mineralization route that converts flue gas CO_2 to precipitated calcium carbonate (PCC) using waste aqueous streams as the calcium source. The proposed mineralization route integrates CO_2 utilization and waste liquid management at coal power plants. Maintaining alkalinity during carbonation is the primary challenge in this process. Instead of relying on the consumption of costly and unsustainable sources of alkalinity (e.g., sodium hydroxide), we used regenerable solids to increase the pH of water by ion exchange. The increase in alkalinity creates a more thermodynamically favorable environment for calcite precipitation.

The capacities for ion exchange of various commercially available zeolites and ion exchange resins in CO_2-saturated water were studied through batch equilibrium and column ion-exchange experiments. During ion exchange, H^+ in the water exchanges with Na^+, Ca^{2+} or K^+ in the zeolite or resin. Batch equilibrium isotherms for H^+–Na^+ exchange were collected by first adding varying amounts of the ion exchange solid (0 – 1.0 kg per kg of water) to CO_2-saturated water (pCO_2 = 1 atm, pH 4) to determine the maximum H^+–Na^+ exchange achievable. Ion-exchange column studies were performed to determine breakthrough capacity of the system. In both systems, an increase in pH coupled with an increase in Na concentration was observed, demonstrating the feasibility of forward ion exchange. Zeolitic materials increased the alkalinity of the solution from pH 4 to around 10, whereas ion exchange resins containing weakly acidic functional groups (e.g., carboxylic acid) exhibited higher affinity for H^+, increasing pH to 11.3 and surpassing the pH shift induced by the zeolitic materials.

The extent of calcite precipitation was evaluated by mixing the alkaline CO_3^{2-}-rich water obtained after passing through the ion-exchange column and a $CaCl_2$ solution. Calcite yields are as expected from chemical thermodynamics. Calcium carbonate yields found experimentally were about 26 and 8 mmol per kg water for the weakly acidic ion exchange resin and zeolitic materials, respectively.

The results from these studies indicate that ion exchange processes can be used as an alternative to the addition of stoichiometric inorganic bases (e.g., sodium hydroxide) to induce alkalinity for the consequent precipitation of $CaCO_3$. Instead, weakly acidic ion exchange resins commonly used for water hardness removal were alternatively used to induce pH swings in the aqueous solutions. The high calcium carbonate yields obtained for the materials examined demonstrate the potential application of this process. In addition, the successful operation of the calcium carbonate production process at standard temperature and pressure conditions support its potential for industrial implementation.

576070 Concentrating Solar Power Based Conversion of Carbon Dioxide to Fuels

We are employing a range of in-house developed technologies to address conversion of impure CO_2-rich waste streams into products that will function as drop-in replacements for petroleum or petroleum fractions. Advantages of PCI's approach include energy efficiencies and improved reactor performances, in terms of both yields and energy efficiencies that are inherent to our reactor designs. The underlying improvements are due to use of Microlith®-based catalyst substrates; use of which enables the demonstrated improved performances as well as enabling reactor designs effective over a wide range of thermal inputs from low kW to high MW scales. This reactor technology enables meeting DOE performance goals of CO_2 conversion, energy efficiencies, and product yields. Additionally, the use of PCI's processes enables the conversion of waste CO_2, without the need for further purification, into value-added saleable products or serves as a drop-in replacement to petroleum or crude oil. We also address the impact of solar or wind variabilities on the operability of our process.

Our approach to upgrading CO_2-rich streams containing CO, CH_4, H_2 and N_2 is to use high-temperature reforming, with the addition of supplemental CH_4 and process recycle, to form syngas. This can be followed by water-gas shift, if needed to adjust the CO/H_2 ratio, then Fischer-Tropsch synthesis to produce liquid-range hydrocarbons for sales or as alternative refinery feedstocks. This process makes use of concentrating solar power (CSP) to provide the high temperatures needed for the reforming reaction, ~850-925 °C. We have demonstrated moderate durability with no evidence of catalyst deactivation or carbon formation, achieving close to equilibrium CO_2 and CH_4 conversions. Our reactor design also minimizes the impact of the significant endotherm that other approaches to CO_2/CH_4 reforming suffer from; our ability to maintain near-optimum temperature through the reactor enables the noted equilibrium conversions while preventing carbon deposition.

Concentrating Solar Power (CSP) technologies, either cone or mirror-illuminate receivers, depending on the required power levels ranging from 250 kW to over 10MW per solar receiver, are used to drive the highly endothermic CO_2/CH_4 reforming process. We estimated that over 50% of solar energy can be converted into fuel energy, exceeding the DOE target of 40% thermal efficiency and at a cost competitive basis with current pump-diesel value.

Microlith mesh is a catalyst substrate consisting of thin catalytically-coated high-alloy metal mesh screens with very small channel diameters. Reactors employing this coated mesh have excellent fluid mixing, elimination of internal thermal gradients, high volumetric concentrations of catalysts, and low pressure drops, which together enable the CO_2 conversion efficiencies observed experimentally. Additionally, the resulting high heat and mass transfers on the Microlith-based reactor allows for compact reactor sizes needed for tight integration with CSP. The open structure leads to extremely low pressure drops, as compared to pellet packed beds, and less of a pressure drop as compared to structured substrates such as honeycomb monoliths. The same structure is also useful for supporting sorbents for capture of CO_2 from flue gas and other sources in addition to use in reactors conducting highly exothermic or endothermic reactions.

578937 Synergy of Biological and Chemical Processes for High-Efficiency CO_2 Capture and Utilization

The power industry in the U.S. emits more than 1.8 gigaton of CO2 (30% of the total CO2 emissions) per year. New solutions are urgently needed to capture and utilize the CO2. An integrated solution including algal cultivation, chemical CO2 absorption, and polymer production was developed to synergistically combine biological and chemical processes with the power plant operation to efficiently capture CO2 and produce value-added chemicals. A pilot system has been established at the Michigan State University power plant, which has been running for more than a year to develop the combined biological and chemical solution. The results from pilot operations were used to carry out a system design. A detailed techno-economic analysis was then conducted on the selected system to conclude a feasible solution for the power industry.

568016 Microwave Active Conductive Metal Oxides for CO_2 Dry Reforming of Methane

The dry reforming of methane reaction (DRM) produces H_2 and CO (syngas) from CO_2 and CH_4. In this regard, DRM enables "carbon neutral" syngas production because it consumes CO_2 as a reactant, whereas conventional methane steam reforming generates over nine tons of CO_2 for each ton of H_2. Using CO_2 to produce commodity chemicals, such as H_2 and CO, can generate revenue to offset carbon capture costs, reduce the carbon footprint of fossil-fuel powered processes, and allow sustainable use of our country's fossil fuel resources. Unfortunately, the high operating temperatures of DRM (>860°C) introduce challenges compared with the lower-temperature, carbon-positive, methane steam reforming.

Recently, microwave-assisted catalysis has been proposed as an enabling technology for high temperature chemical processes. Unlike traditional thermal heating, microwaves can rapidly heat catalysts to extremely high temperatures, which allows reactors to utilize excess renewable energy on an intermittent basis (load follow) to promote traditionally challenging, thermally-driven reactions. Moreover, microwaves directly heat the catalyst material, and not the entire reactor body, which reduces many of the traditional heat management challenges associated with high temperature DRM. Microwave absorption is a function of the electronic and magnetic properties of a material, and a properly designed catalyst can function as a both a microwave heater and a reactive surface for driving the desired reaction. However, microwave absorption is extremely sensitive to the catalyst's chemical state and electronic structure, and effective catalysts must maintain microwave activity across a wide range of temperatures and in both oxidative and reductive environments.

We have constructed a single-mode microwave reactor and investigated microwave assisted DRM using conductive metal oxides. Conductive perovskites (ex: $La_{0.8}Sr_{0.2}CoO_3$) are attractive because they readily convert microwave energy into heat while also potentially providing active metal and oxide surfaces for catalyzing the reaction. Additionally, perovskites have good thermal stability at temperatures in excess of 1000°C, unlike conventional carbon-based microwave absorbers which must be kept below 900°C to avoid pyrolytic carbon sublimation. The conductivity of these oxides allows them to inductively heat in the microwave like metals or carbon-based absorbers, but with the advantage of maintaining conductivity even in oxidative environments. We have demonstrated the rapid heating of the microwave DRM reactor and were able to ramp from a ground state to reaction temperature (>700°C) in less than 3 minutes with consistent on-off cyclability. By modifying the transition metal dopants used at the perovskite B site, we were able to tune the catalyst performance such that stable conversions of 90% of CO_2 and CH_4 could be achieved using less than 100W of applied microwave power, demonstrating the suitability of microwave catalysts for rapid, efficient, and carbon neutral syngas production.

554205 Nano-Engineered Catalyst for the Utilization of CO_2 in Dry Reforming to Produce Syngas

The objective of this study was to develop a novel catalytic reactor containing nano-engineered catalyst for the utilization of CO_2 (captured from coal-fired power plants) in dry reforming of methane (DRM) ($CO_2 + CH_4 \rightarrow 2\ H_2 + 2\ CO$) to produce synthesis gas (syngas). The technology aims to reduce CO_2 emissions by developing beneficial uses for CO_2 from coal-fired power plants. It also offers an alternative to mitigate CO_2 emissions in areas where geologic storage may not be an optimal solution and/or utilization could significantly offset the costs of carbon capture and sequestration.

The nano-engineered Ni-based catalyst was prepared by atomic layer deposition (ALD). The Ni particles were as small as ~2-4 nm. The nano-engineered catalyst showed CH_4 conversion >95%, H_2/CO ratio in the range of 0.7-1.0, and CH_4 reforming rate as high as 2,500 $L/h/g_{Ni}$ at 850 °C and pressure of 15-25 psia. The Ni-based ALD catalyst also showed good stability in DRM reaction during a 200-hour continuous operation at 850°C. This is due to a strong bonding between the nanoparticles and substrates since the Ni nanoparticles were chemically bonded to the substrate during the ALD process. The high thermal stability maintains high dispersion of Ni nanoparticles, which can inhibit coke formation, because their step edges are small enough to limit carbon nucleation and growth.

572276 Tale of an Ethane Oxidative Dehydrogenation Catalyst with Long and Stable Life Under Harsh Conditions

Southern Research, in partnership with DOE/NETL has developed a novel, low-cost mixed oxide catalyst for moderate temperature ethylene production process using CO_2 and ethane. The proprietary catalyst exploits CO_2 as a 'soft' oxidant to dehydrogenate ethane to ethylene at high selectivity (oxidative dehydrogenation). Long term stability of catalyst has been studied using simulated coal-fired flue gas (as CO_2 source). This testing was conducted for up to 500 hrs which constituted 50+ cycles of reaction followed by regeneration. The catalyst showed remarkable activity, selectivity, stability and regenerability at harsh conditions in presence of some notable coal fired flue gas impurities. Techno-economic (TEA) study shows the cost of ethylene production can be competitive with partial co-product credit. On the other hand, life cycle assessment (LCA) reveals potential for 60-70% reduction in CO_2 emission for an integrated power-ethylene plant compared to the state-of-the-art alternatives. The catalyst synthesis has been scaled up to advance the technology from laboratory scale to field scale demonstration. The presentation will highlight the results from the long term testing study, ongoing efforts to scale up process and the updated TEA/LCA results.

585305 Farming Algae for Animal Feed with CO_2 Utilization

Photosynthetic production of microalgal biomass production for fuels, feeds, foods and other low-cost commodities will also require low-cost sources of CO2. Unlike higher plants, microalgae have low access to atmospheric CO2 due to the low mass transfer rate from $CO_{2(gas)}$ to $CO_{2(aqueous)}$, resulting in limited algae growth. Thus, most research in this field has focused on concentrated CO2 sources, in particular CO2 from the flue gases of fossil-fueled power plants. However, the transport of the flue gases (containing from about 3-13% CO2) from source to the algal production site must be limited to a few kilometers due to cost. The alternative for longer distance transport - also expensive - is to concentrate the flue gas CO2 by chemical processes. More concentrated, sources of CO2 are also available, such as from refineries, fertilizer and other chemical plants, ethanol fermentations, anaerobic digestion of wastes, among others. Microalgae can be cultivated on organic wastes, which provide both nutrients (N, P, K, etc.) and organic and inorganic carbon for algal growth. Finally, direct capture of CO2 from air by the algal cultivation system may be possible, or it could be coupled to an air-capture process. The relative productivities, economics, greenhouse gas balances, and resource potentials of these alternative CO2 sources for large-scale production of microalgae are reviewed. A case example will be presented for a currently ongoing project for utilization of CO2 from a coal-fired power plant in Orlando Florida, for production of biofuels, animal feeds and wastewater treatment. Annual productivities of filamentous algae and their feed nutritional value will be presented.

573613 Modular Electrocatalytic Processing for Simultaneous Carbon Utilization and Alkane Conversion

U.S. coal plants play an important role in providing the U.S. with reliable low-cost electricity and will continue to be relied upon as base-load for the foreseeable future. Cost-effectively reducing carbon emissions from the existing U.S. coal-fired generation fleet is necessary for these assets to continue to play an important role in a carbon restricted economy. Carbon utilization is a methodology which could help offset the costs of carbon capture by converting the product CO_2 into valuable products.

Carbon Monoxide (CO) is an important industrial gas used in manufacturing bulk chemical precursors such as phosgene, commodity materials via carbonylation including aldehydes, ketones, carboxylic acids, anhydrides, esters, amides, imides, carbonates, ureas, and isocynanates. Further, high purity CO (>99.99%) is used in electronics manufacturing. Industrial bulk CO is produced by separating CO from syngas (containing H_2) generated via steam methane reforming (SMR) using natural gas as the feedstock. The U.S. consumes approximately 300-400 MMscf/day of CO with an annual market value of nearly $2 billion.

Various separation technologies are used including cryogenic separation (i.e. cold box), pressure swing adsorption (PSA), membrane separation, and ammonium salt solution absorption. However, these processes are complex and capital intensive requiring larger production facilities. Although low temperature solution-based CO_2 electrolyzers are being studied, solid oxide electrolyzer cells (SOECs) operating at an elevated temperature provide greater cell/electrolysis efficiencies. The ability to utilize solid oxide electrolyzer cell (SOEC) technology to electrochemically convert CO_2 into CO/O_2 or CO_2/H_2O into syngas/O_2 has been shown over recent years [1-3].

Ohio University (OHIO) with funding from the U.S. Department of Energy's National Energy Technology Laboratory (NETL) Carbon Use and Reuse Program (DE-FE0031709) is developing a process that utilizes solid oxide electrolysis to simultaneously convert CO_2 and ethane contained in wet natural gas into valuable products. OHIO's SOEC process seeks to advance beyond these previous electrochemical CO_2 reduction concepts by utilizing intermediate-temperature operation (650-750°C) and utilizing an innovative cell design which incorporates oxidative dehydrogenation (ODH) of C_{2+} alkanes. Several transition metals and mixed metal oxides have been identified as suitable for solid oxide electrolysis of CO_2 and ODH of C_{2+} alkanes using first principles simulations [4] and experimental trials, respectively. This presentation will discuss recent results from OHIO's carbon utilization project including experimental results and process simulations.

[1] F. Bidrawn, G. Kim, G. Corre, J. T. S. Irvine, J. M. Vohs, and R. J. Gorte, "Efficient Reduction of CO2 in a Solid Oxide Electrolyzer," Electrochem. Solid-State Lett., vol. 11, no. 9, pp. B167–B170, Sep. 2008.

[2] L. Zhang, S. Hu, X. Zhu, and W. Yang, "Electrochemical reduction of CO2 in solid oxide electrolysis cells," J. Energy Chem., vol. 26, no. 4, pp. 593–601, Jul. 2017.

[3] Y. Xie, J. Xiao, D. Liu, J. Liu, and C. Yang, "Electrolysis of Carbon Dioxide in a Solid Oxide Electrolyzer with Silver-Gadolinium-Doped Ceria Cathode," J. Electrochem. Soc., vol. 162, no. 4, pp. F397–F402, Jan. 2015.

[4] X.-K. Gu, J. S. A. Carneiro, and E. Nikolla, "First-Principles Study of High Temperature CO2 Electrolysis on Transition Metal Electrocatalysts," Ind. Eng. Chem. Res., vol. 56, no. 21, pp. 6155–6163, May 2017

565690 Process Intensification for Electrochemical Utilization of CO_2

The conversion of CO_2 to value-added products is an attractive approach for reducing the cost of CO_2 capture. Formic acid (FA) is a multipurpose industrial chemical feedstock for formic acid fuel cells, additives in the textile and agricultural sectors, cleaning products, and corrosion inhibitors. FA can be produced directly from CO_2 via a two-electron process facilitated with DC electricity supplied via renewable/intermittent power sources [1]. The state-of-the-art method for industrial FA production is the Kemira process, which uses methanol, sodium methoxide, and CO to produce methyl formate that is subsequently hydrolyzed to formic acid.

Conversely, an electrochemical approach reduces the number of chemical input for FA production, but conventional electrochemical processes for CO_2 reduction to FA continue to suffer from selectivity issues where the electrocatalyst produces a wide range of products. These competing products include FA, hydrogen, carbon monoxide, methanol, and methane that reduce faradaic efficiency by up to 50% [2]. The CAER-Andora process features an enzymatic catalyst with high selectivity for FA production with turn-over-frequency >85 s^{-1}. The CAER-Andora process combines an electrochemical cell that features graphite electrodes and an electrochemically facile charge mediator, combined with an immobilized proprietary catalyst inside a FA production cell. These processes work together to 1) reduce faradaic inefficiencies by producing only FA, 2) protect the valuable catalyst from deactivation by overpotentials, and 3) provide a modular system configuration that facilitates independent optimization of the electrochemical and production cells. Operating at <3 V, the CAER-Andora process currently produces ~1 mM FA/hour with >90% CO_2 to FA conversion in the production cell, and progress towards the intensification of the CO_2 utilization process will be discussed.

References

[1] Yang, N., S.R. Waldvogel, and X. Jiang, *Electrochemistry of Carbon Dioxide on Carbon Electrodes.* ACS Applied Materials & Interfaces, 2016. **8**(42): p. 28357-28371.

[2] Huo, S., et al., *Coupled Metal/Oxide Catalysts with Tunable Product Selectivity for Electrocatalytic CO2 Reduction.* ACS Applied Materials & Interfaces, 2017. **9**(34): p. 28519-28526.

556289 Integrated Process of CO_2 Capture and Conversion to Alkyl Carbonates

A new generation of CO_2-capture and utilization technologies that are being separately developed by federally funded projects would show market-driven adoption if CO_2 were used to manufacture value-added products competitively. Dimethyl carbonate (DMC), one such product, has an expanding world market: a) for polycarbonate production; b) as electrolyte in lithium-ion batteries; c) as intermediate in polyurethane production; and d) as environmentally friendly solvent replacing ketone-based solvents. DMC also is being evaluated as an additive to reduce emissions from diesel engines.

Integrating the process of CO_2 capture and conversion to DMC should significantly enhance both the economics as well as energy efficiency when compared to separate processes for CO_2 capture followed by conversion. The direct reaction of CO_2 with methanol for synthesis of DMC is thermodynamically difficult, requiring innovative catalyst development. Therefore, E3Tec is actively pursuing two separate integrated processes based on chemical carriers for CO_2 utilization to manufacture high-value DMC; one with the SBIR Phase II (DE-SC0013233) and the other one with an ERA, Alberta grant. The two DMC processes are based on two alternate chemical pathways 1) using ammonia as a carrier while consuming CO_2 and methanol feeds; and 2) using CO_2, methanol and ethylene oxide (EtO) as feeds to co-produce DMC and mono-ethylene glycol (MEG).

E3Tec is now pursuing integration of the DMC process with this new generation of CO_2 capture technologies, i.e., membrane separations; solid adsorbents; non-aqueous solvents; a new generation of amine solvents; and hybrid processes. The technology challenges and opportunities of integrating the DMC process with the new generation of CO_2 capture technologies are discussed with the focus on different membrane separations and solid adsorbents. The ASPEN Plus® analysis of the integrated process is based on prototype testing of the DMC process and reported performance of the CO_2 capture processes.

554488 Electrochemical CO_2 Conversion to Valuable Chemicals

Converting greenhouse gas carbon dioxide (CO_2) to value-added chemicals is an appealing approach to tackle CO_2 emission challenges. The chemical transformation of CO_2 requires suitable catalysts that can lower the activation energy barrier to minimize the energy penalty associated with the CO_2 reduction reaction. Cu is the only monometallic catalyst that can produce an appreciable amount of C2+ products from CO2, while the C2+ selectivity of Cu must be further improved in order to be considered for commercialization. Here, we will present our work in developing an integrated two-stage electrolyzer stack system for the conversion of CO_2 to alcohols. The first stage utilizes a silver catalyst in a flow cell capable of reducing CO_2 to CO with faradaic efficiencies of >95% and partial current densities of >200 mA/cm^2. For the second stage, a CO reduction flow cell using alkaline electrolyte can reduce CO to C2+ products with high selectivity and high current density. We recently constructed a high-performance CO flow electrolyzer with a well-controlled electrode-electrolyte interface that can reach total current densities up to 1 A/cm^2 together with improved C_{2+} selectivities. Computational transport modelling and isotopic $C^{18}O$ reduction experiments suggest that the enhanced activity is due to a higher surface pH under CO reduction conditions, which facilitated the production of acetate. At optimal operating conditions, we achieved a C_{2+} Faradaic efficiency of ~91% with a C_{2+} partial current density over 630 mA/cm^2. Further investigations show that maintaining an efficient triple-phase boundary at the electrode-electrolyte interface is the most critical challenge to achieving a stable CO/CO_2 electrolysis process at high rates.

561175 Highly Active, Robust 3D Interconnected Porous Oxide-Derived Copper Electrocatalysts for Selective Conversion of CO_2 into Fuels

The production of value-added chemicals and fuels from CO_2 and water by highly efficient, selective, robust electrocatalytic materials is of vital interest to address the energy challenges. Expensive noble metal catalysts have been extensively studied for electrochemically converting CO_2 into CO with high selectivity at low overpotentials. However, there is also a need to abate and/or replace the use of precious metals with Earth-abundant transition metals. In this work, we report hierarchical CuO-derived inverse opal catalysts which demonstrated impressive CO selectivity compared with commercially-available CuO powders and other reported copper-based catalysts. Such a 3D interconnected porous catalyst composed of small nanoparticles surprisingly produced CO with a peak Faradaic efficiency (FE) of 72.5% at -0.6 V *vs.* RHE and was highly robust during 24-hour CO_2 electrolysis. Selective C_1 production alongside negligible H_2 evolution could be ascribed to a local pH increment inside the interconnected porous structure and rapid consumption of CO_2 and protons at the highly roughened catalyst surface during the electrolysis. In addition, the *in situ* synchrotron X-ray absorption fine structure and X-ray diffraction measurements allowed us to probe the reduction of CuO into metallic Cu with dominant active Cu (111) surfaces under electrocatalytic conditions. The CO_2-to-CO outperformance of our 3D porous oxide-derived copper materials over typical oxidized copper catalysts could be ascribed to the synergistic effect of roughened, interconnected inverse opal morphology and surface structure which would highly correlate with weak binding of the intermediate species, lower adsorbate mobility, and reduced reactant availability per active site. Our anticipated findings may open an opportunity for rational design of promising, high-performance, less expensive electrocatalysts to produce sustainable fuels and value-added chemicals from fossil fuel-generated CO_2 emission.

579543 NO_X Emissions for Rotating Detonation Combustion Operating on Hydrogen-Air and Hydrogen/Natural Gas - Air

Pressure Gain Combustion (PGC) offers a means of improving gas turbine thermal efficiency that can be combined with, and thus surpass gains available from higher turbine inlet temperatures alone. Conventional gas turbine combustion relies on deflagration in an idealized constant pressure combustion process that in reality is subject to a pressure loss. However, PGC achieved through detonation, provides an alternative pathway to similar combustion temperatures but with less entropy generation, and because of the gain in pressure, more work availability. Utilizing a Rotating Detonation Combustor (RDC), it is possible to achieve a near continuous detonation wave that propagates circumferentially in an annular combustor with bulk axial flow that is conducive to a conventional turbine design. In addition to high thermal efficiency, industrial and power generation turbines burning coal syngas in air or natural gas in air must also adhere to strict oxide of nitrogen (NOx) emissions.

The focus of this study is to explore, both experimentally and numerically, NOx emissions from an RDC operating on hydrogen (coal syngas) in air and natural gas/hydrogen in air. Due to short residence times within the detonation wave and rapid expansion downstream of the wave, the production of NOx has the potential to be less than a comparable non-premixed, turbulent deflagrative flame. Tests performed in an uncooled RDC with an outer diameter of 152 mm and an annular gap of 7.62 mm were conducted over a range of equivalence ratios (0.7 – 1.0) while collecting NOx and O2 emissions at multiple locations in the exhaust stream. As variations in equivalence ratio induce changes in the modes of wave propagation, it is possible to consider the impact of single versus multiple wave operation on emissions. Experimental results are compared to predicted NOx values attained through a Chemical Reactor Network (CRN) derived from a Computational Fluid Dynamics (CFD) model utilizing chemical kinetics mechanisms tuned for detonation.

578131 Ultrasonic Measurements of Temperature Distribution and Heat Flux Vectors in Solid Components of Energy Conversion Processes

Ultrasonic Measurements of Temperature Distribution and Heat Flux Vectors in Solid Components of Energy Conversion Processes

Yunlu Jia and Mikhail Skliar, University of Utah, Salt Lake City, UT

Background, Motivation, and Objective: Extreme environmental conditions of high temperatures, pressures, chemical aggressiveness, and mechanical abrasion are common in military, airspace, energy conversion, and nuclear applications. In such environments, the most hardened insertion sensors do not perform reliably for long. For such environments, we have developed an ultrasonic method for measuring the spatial distribution of temperatures in solid materials and, specifically, across containments of extreme processes.

Statement of Contribution/Method: We describe a novel ultrasonic approach to the measurements of the temperature distribution and heat fluxes in solids [1-3]. We use the ultrasound propagation paths or waveguides structured to contain engineered or naturally occurring echogenic features which produce a train of echoes in response to an excitation pulse, Fig. 1. The time of flight between echoes encodes the information on the temperature distribution in the corresponding segment, which we reconstruct using its parametrization by the heat conduction model or as a piecewise linear function.

Results: This paper describes the laboratory and pilot-scale testing of the developed approach. The validation results show that the estimated temperature profile is correctly captured, and the measurement accuracy can be comparable with traditional insertion sensors. It is further shown that by supplementing the ultrasound data with the surface temperature measurements, the entire volumetric distribution of the internal temperature inside of solid samples can be estimated noninvasively and used to reconstruct the corresponding internal heat fluxes. We will discuss the advantages this approach, which include its suitability for in-situ and non-destructive measurements in extreme environments, a broad range of temperatures, locations inaccessible to the traditional insertion sensors, and the applicability to the measurements of heat fluxes in solids.

[1] Y. Jia & M. Skliar, *Energy & Fuels*, 2016. [2] Y. Jia et al., *Ultrasonics*, 2016. [3] M. Skliar, et al., US Patents 8,801,277 B2 and 9,212,956.

562146 CFD Simulations of Propane/Natural Gas Blended Fuels Combustion in Gas Turbine

Changes to the US electricity industry are forcing a paradigm shift in how the nation's generating assets are operated. Coal-fired power plants optimized as baseload resources are being increasingly relied on as load-following resources to support electricity generated from intermittent renewable capacity. Existing plants are being operated in ways that are suboptimal from the perspective of efficiency and capacity utilization. In an effort to address these challenges by identifying and developing technologies that increase coal-fired power plant efficiency, improve unit reliability and availability, and enhance unit capability for flexible operations (e.g. “cycling”), a CFD model is developed to simulate the propane/natural gas blend combustion rig.

Unconventional gas supplies, like shale gas, are expected to grow which will make U.S. natural gas composition more variable and the composition of fuel sources may vary significantly from existing domestic natural gas supplies. The effect of gas composition on combustion behavior is of interest to allow end-use equipment to accommodate the widest possible gas composition. The CFD model is first validated with the available experimental data which have been collected from a pressurized single injector combustion test rig at the National Energy Technology Laboratory (NETL). The tests are conducted at 7.5 atm with a 589K air preheat. A propane blending facility is used to vary the site natural gas composition. The CFD model predicted results are within the experimental error bar. After the CFD model is validated with experimental results, then a much wider range of gas composition is simulated and the effect of gas composition on combustion is analyzed and discussed.

561931 Advanced Oxy-Combustor Concepts for Direct-Fired Supercritical CO_2 Power Cycles

The supercritical CO2 (sCO_2) power cycle concept is identified as a potentially efficient, economical, and pollutant free power generation technique for future power generation. Recent work in the literature provides some strategies and best operating conditions for direct-fired sCO_2 combustors based on zero-dimensional reactor modeling analysis, however there is a need for a detailed investigation using accurate combustion chemical kinetics and thermophysical models. Here, the sCO_2 combustor is modelled by coupling perfectly stirred reactor (PSR) and plug flow reactor (PFR) models. The real gas effects are incorporated using the Soave-Redlich-Kwong (SRK) equation of state. Also, the detailed Aramco 2.0 kinetic mechanism is used for the combustion kinetic rates. It is found that the primary zone must be diluted either with thirty or forty-five percent of the total CO_2 in the cycle to have a feasible combustor design. However, the forty-five percent dilution level at 950 K and 1000 K yielded a better consumption of CO, O_2 and CH_4. Also, the cross-sectional area of the sCO_2 combustor can be scaled-down to 10 to 20 times smaller than a traditional combustor with the same power output. Further, from this investigation, it is also recommended to have a gradually increasing secondary dilution in the dilution zone, by using progressively larger diameter holes. This design would help retain relatively high temperature in the initial portion of the dilution zone and would help consume fuel species such as, CO and CH_4. It appears that, for sCO_2 combustors "lean burn" is the better strategy over stoichiometric burning to eliminate CO build up at the combustor exit. The lean burn condition at equivalence ratio, ϕ, =0.9 is recommended for sCO_2 combustor operation. Also, the length of the dilution zone can be scaled-down to 50% by lean burn operation of the combustor. It is also observed that the lean burn increases the net turbine power. Computational Fluid Dynamic simulations of the direct-fired systems identified using 0-D modeling will also be presented in the work. Current work provides crucial design considerations for the development of advanced sCO_2 combustors to be used with direct-fired power cycles.

568338 Pressurized Oxy-Fuel Combustion in Fluidized Bed: 3D CFD Simulation with Experiment

Abstract

For reducing the global greenhouse gas emissions and slowing global warming from the process of industrialization, carbon capture and storage (CCS) technologies have been supposed to reduce carbon dioxide emission apply from the power plants and fossil industries. Oxy-fuel combustion technology has been considered as a promising means to achieve the objective of effective CO_2 capture with commercial potential. The fluidized bed with oxy-fuel combustion technology can fire coal and low quality fuels like solid wastes and biomass with high efficiency and lower pollutants emissions. However, the net efficiency will fall by more than 10% primarily caused by the high energy consumption of air separation unit (ASU) and the compression purification unit (CPU) and the pressure mismatch among the ASU, furnace and CPU. Pressurized oxy-fuel combustion (POFC) technology as a new generation of oxy-fuel combustion technology is a perspective application in industrial process due to its higher net efficiency and combustion efficiency than ordinary oxy-fuel combustion.

Understanding the detail gas-solid flow and combustion characteristics is of paramount importance to the design, operation and optimization of oxy-fuel FB reactors. In this work, a 3D Eulerian-Lagrangian model including particle flows and chemical reactions was developed to simulate the pressurized oxy¨Cfuel combustion process in FB (Fig. 1) based on the Multi-Phase Particle-In-Cell (MP-PIC) scheme, and the particle size distribution was particularly taken into account. As shown Fig. 2, this model successfully predicts critical parameters by comparing with the experimental data from a pilot-scale 15 kW_{th} pressurized fluidized bed. On this foundation, some detailed combustion characteristics and full-physics picture including particle flow behavior that are hardly measured by experiment were simulated. The simulated results show that the intensity of particle motion and expansion degree of bed was gradually decreased with an increase in pressure under same gas flow rate. The higher operating pressure led to higher particle concentrations in the furnace and caused the variation of flow pattern. the temperature of solid phase region gradually increased with pressure and inlet O_2 concentration increased. In addition, the elevated pressure and inlet O_2 concentration can help to reduce the emission of CO and NO_x. In addition, the size and size distribution of coal in FB were also study in this work with different operating pressure and inlet oxygen concentration. This work can be expected to improve and optimize the scale-up pressurized oxy-fuel FB system for industrial application in future research.

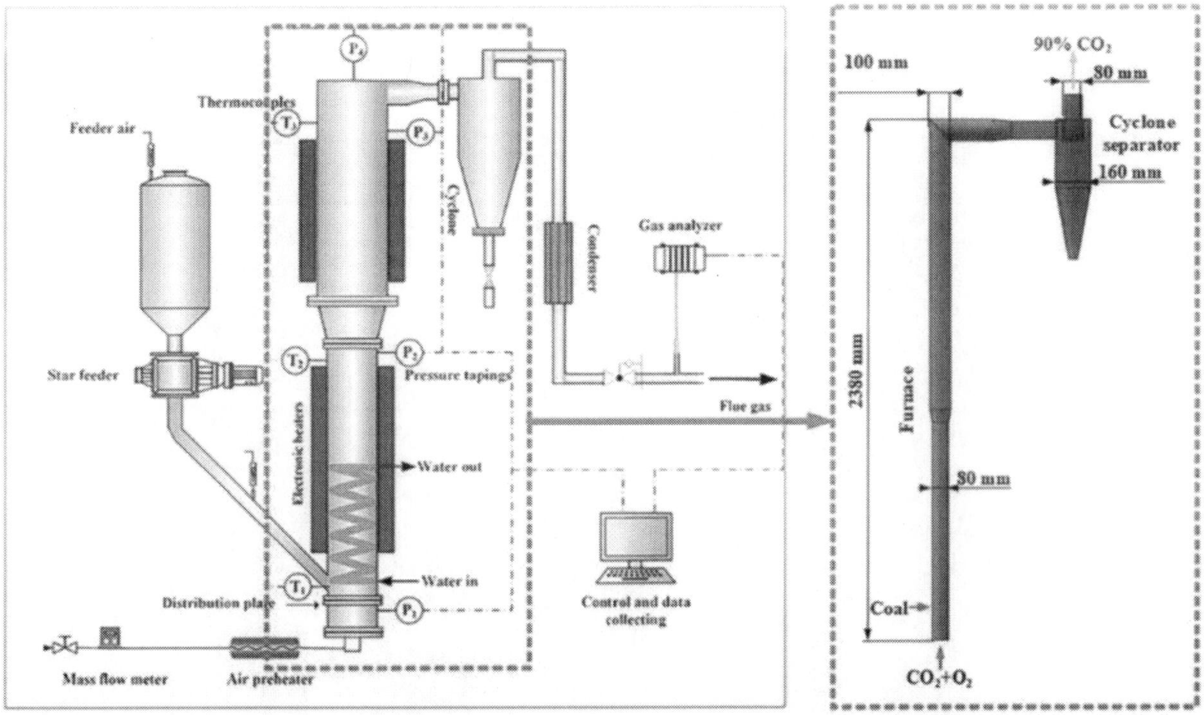

Fig. 1. Three-dimensional CFB model simplified from the 15 kW_{th} pilot-scale pressurized fluidized bed

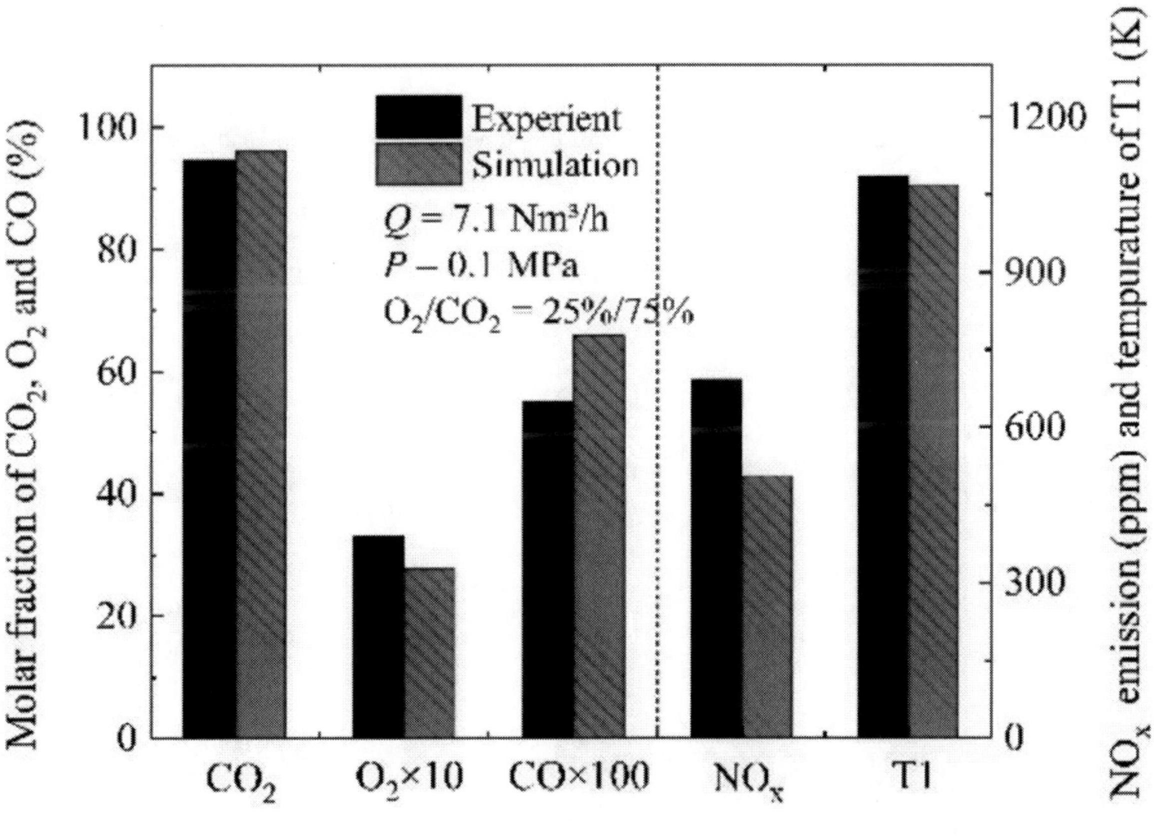

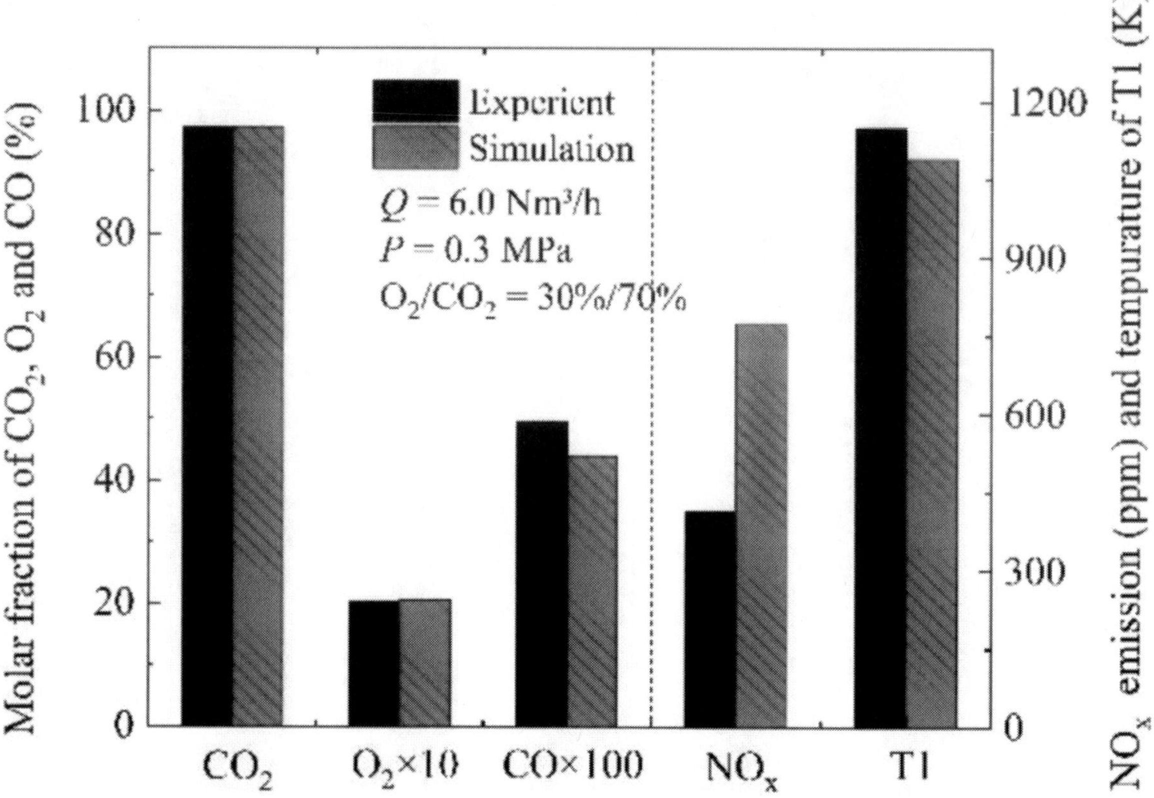

(a)

(b)

Fig. 2. Comparison of the predicted results and the experimental data under different conditions

558857 Experimental Study of so$_2$ Emissions and Desulfurization of Oxy-Coal Combustion in a Pressurized Fluidized Bed Combustor

Experimental study of SO_2 emissions and desulfurization of oxy-coal combustion in a 30 kW_{th} pressurized fluidized bed combustor

Lei Pang[1], Yingjuan Shao[1,*], Wenqi Zhong[1]

1. Key Laboratory of Energy Thermal Conversion and Control of Ministry of Education, School of Energy and Environment, Southeast University, Nanjing 210096, P.R. China

Oxy-fuel combustion has been widely accepted as a highly-effective CO_2 capture and storage (CCS) technology. A combination of this technology and fluidized bed combustion technology, i.e., oxy-fuel-based fluidized bed combustion, had attracted attention from researches during last decade due to its advantages in terms of fuel flexibility, furnace desulfurization and low NO_x emission. Among these studies of oxy-fuel combustion, SO_2 emission is a major concern because it leads to not only the corrosion of the tube and device but also acid rain. In addition, the high SO_2 concentration in flue gas also has negative impacts on CO_2 compression and purification. Although oxy-fuel combustion has shown advantages and potentials in both combustion and CO_2 capture processes, the biggest obstacle to its application is the net efficiency loss associated with the cost of the air separation unit (ASU) and compression purification unit (CPU). It is estimated that the net efficiency can be reduced by more than 10% when a conventional air-fired coal power plant is converted to oxy-firing. In order to improve the efficiency, the pressurized oxy-fuel combustion (POFC) technology has been proposed, and many economic system analyses have proved that an increase in combustion pressure can bring economic benefits as it is helpful to recover more thermal energy from the flue gas and reduce the energy cost of CO_2 compression work. However, there are few experimental researches about POFC technology, and studies of SO_2 emissions and desulfurization under POFC conditions are even less.

This study presents the SO_2 emission behaviors under both air and oxy-coal combustion in a 30

kW_{th} fluidized bed combustor with different pressures, and the effects of combustion atmosphere, pressure and temperature on the desulfurization of limestones were investigated carefully. The results showed that the SO_2 emission in air was 15-20% less than that in oxy-30 atmosphere as more sulfur sunk in ashes, while the increase of combustion pressure from 0.1 MPa to 0.4 MPa had no effect on SO_2 emissions. Combustion temperature can affect the desulfurization of limestones in both atmospheric and pressurized oxy-coal combustion conditions significantly, and the desulfurization efficiency increased by more than 30% when the bed temperature increased from 850 °C to about 950 °C. Although the indirect desulfurization of limestones cannot occur under 0.4 MPa and oxy-30 combustion condition with a bed temperature of 950 °C, its desulfurization efficiency reached about 80% which was close to the value of conventional atmospheric air combustion.

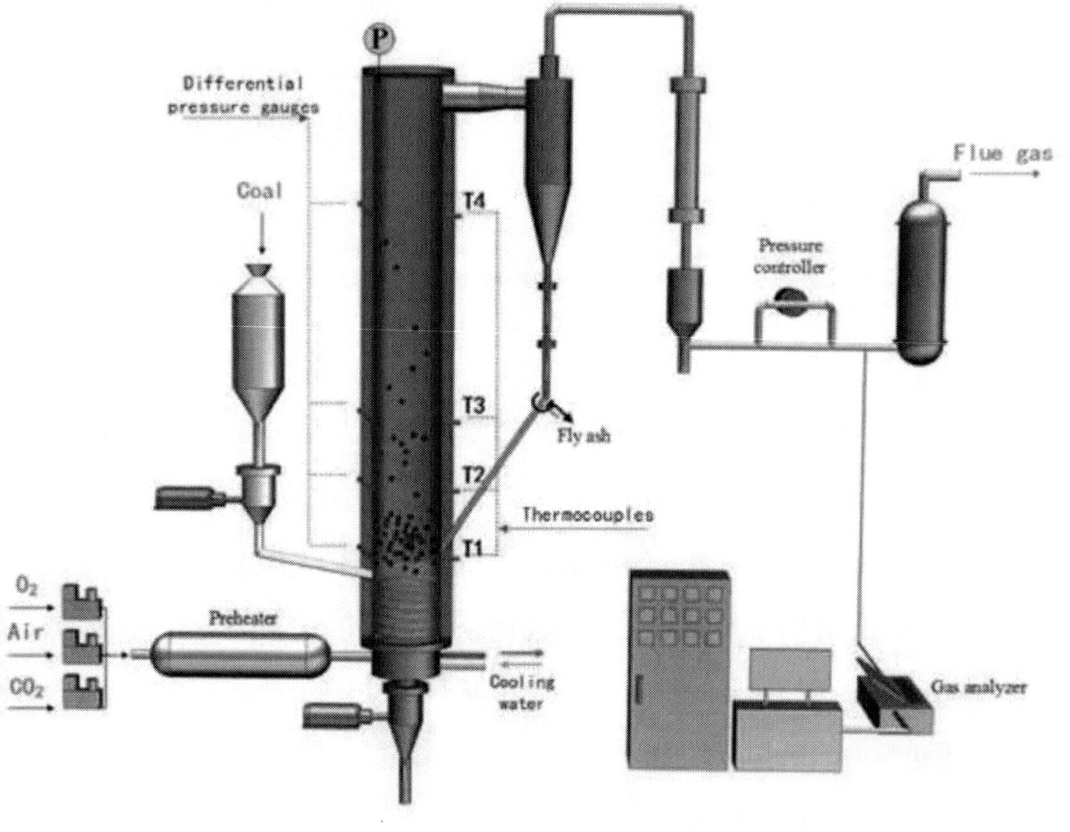

Figure 1. Schematic of the pressurized oxy-coal fluidized bed combustion system

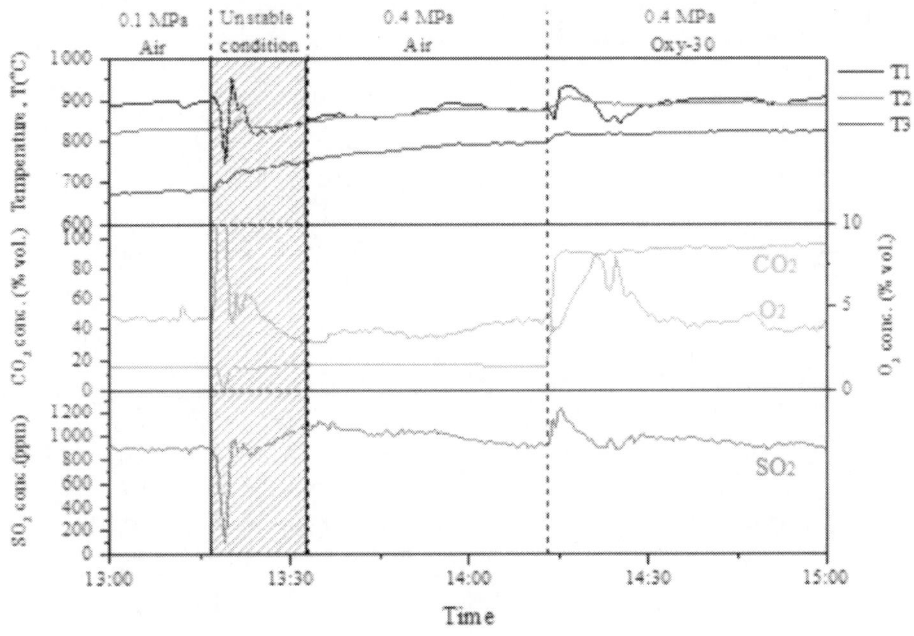

Figure 2. Temperature profiles and gas concentrations at the outlet of the combustor without limestone

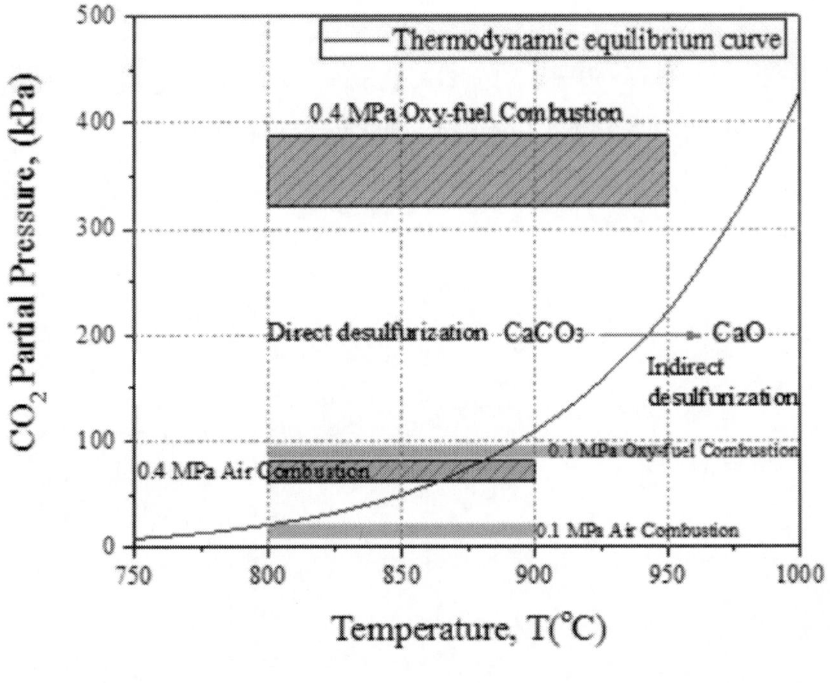

Figure 3. Thermodynamic equilibrium curve of $CaCO_3$ calcination

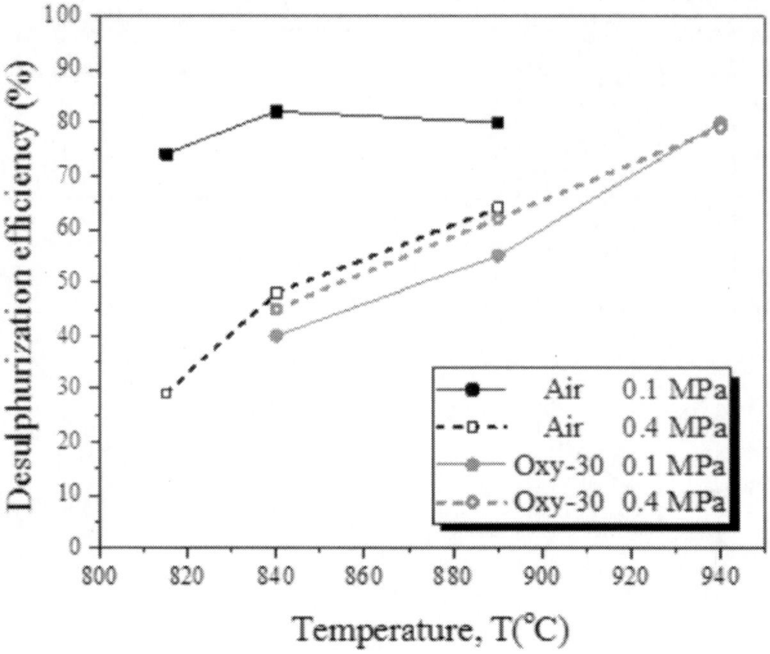

Figure 4. Effect of temperature on desulphurization efficiency

* Corresponding author

E-mail address: yjshao@seu.edu.cn (Y. Shao)

566907 Assessment of Oxy-Fuel Combustion System with Coal and Biomass Co-Firing Based on Circulating Fluidized Bed

Oxy-fuel combustion with circulating fluidized bed (CFB) is one of the most competitive technology for carbon capture and sequestration. Circulating fluidized bed (CFB) combustion as a typically technical direction in oxy-fuel combustion, has several advantages against pulverized boiler, e.g., fuel flexibility, low NOx emission, reduction of recirculated flue gas and higher oxygen concentration. So far, few of the research related to oxy-fuel combustion refers to the system performance with the integration of CFB power plant, air separation unit (ASU) and carbon dioxide compression and purification unit (CPU), especially the system efficiency influenced by different operational parameters. In addition, biomass as an important clean energy source, the system performance of coal and biomass co-firing system combined with oxy-fuel combustion is still unknown.

In this work, a supercritical oxy-fuel combustion system based on CFB boiler firing coal and the mixture of biomass and coal is established according to the conservation of material and energy as shown in Figure 1. Three primary units are arranged in the system including an ASU, a CFB power plant and a CPU. Coal and the blends of coal and biomass burn in the boiler with the oxidants consisting of the oxygen from ASU and the recirculated flue gas. And the combustion process is simplified to pyrolysis, volatile combustion and char combustion. After the validation of adiabatic combustion temperature, the influence of wet/dry cycle, exhaust flue gas temperature and combustion pressure on the net electrical efficiency are evaluated. The excess oxygen ratio and oxygen concentration in the system are determined as 1.06 and 26%, respectively based on the previous research which is satisfied to achieve better performance. The results showed that under the same exhaust flue gas temperature, the flue gas loss of dry cycle is higher than wet cycle mainly because the energy loss of water condensation in dry cycle. Although the co-firing cases have less flue gas loss than coal combustion in wet cycle, the supplement of biomass will decrease the net electrical efficiency because of the pre-treatment of biomass as shown in Figure 2. In addition, a sharp increase occurs when the exhaust flue gas reaches below 60 centigrade because the steam vapor in the flue gas begin to condense. Elevating the combustion pressure is an effective way to improve the heat recovery. The maximum heat recovery at 1MPa is 30MW higher than the atmospheric condition. However, the selection of combustion pressure should also consider the auxiliary energy consumptions of ASU and CPU. In addition, air ingression will significantly influence the net efficiency that 10% of the theoretical air will cause 1% reduction of net electrical efficiency.

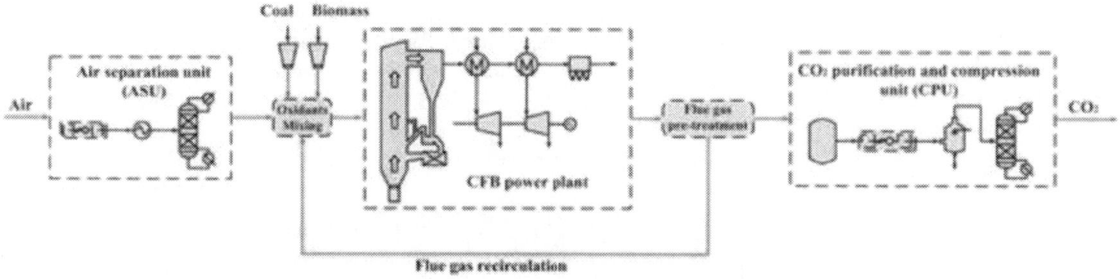

Figure 1. Flow diagram of oxy-fuel combustion power plant based on CFB boiler applied in this work.

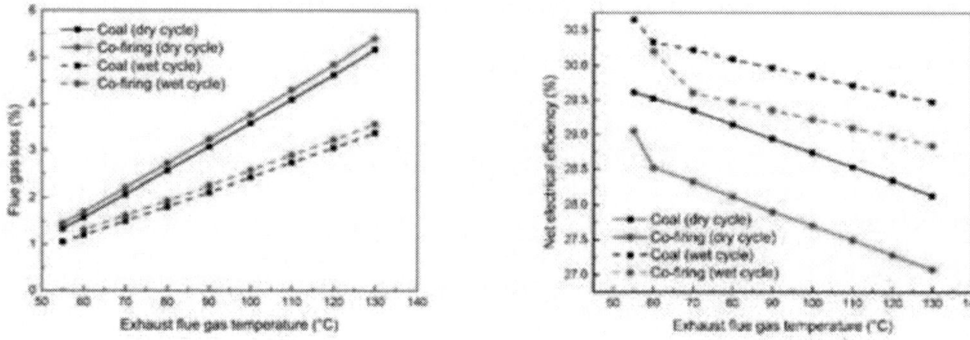

Figure 2. The flue gas loss and net electrical efficiency under varied exhaust flue gas temperature.

577967 Toward a Risk-Based Approach to Justify for Closure at Geologic Carbon Storage Sites

To enable full-scale implementation of geologic carbon storage(GCS) technology stakeholders must have confidence in the ability of GCS sites to safely store large volumes of CO_2 out of the atmosphere for long periods of time, while also protecting human health and the environment. Post-injection site care at GCS sites requires operators to demonstrate non-endangerment of underground sources of drinking water before site closure can be justified. In the study presented here numerical simulation and risk assessment results from a set of hypothetical, but realistic geologic carbon storage sites are considered. These results are used as the basis for defining a science-based, quantitative workflow to justify closure at GCS sites. Using monitoring observations to update performance predictions through the period of active injection period can substantially reduce uncertainties in storage system behavior, that driving force for leakage decreases after injection operations stop, and that in storage sites with reservoirs "stable" it should be safe to reduce post-injection site care requirements. This workflow represents a defensible, science-based approach to evaluate the robustness of GCS site integrity amidst remaining uncertainty, and to justify closure decisions to ensure the long-term protection of groundwater resources.

570127 Investigation of Mass Transfer and Sorption in CO_2/Brine/Rock Systems via *in-situ* FT-IR

CO_2 geological sequestration in deep saline aquifer is considered a promising method to mitigate manmade CO_2 emissions and, thereby, minimize changes to the earth's atmosphere. A fundamental understanding of CO_2 mass transfer and sorption phenomena in brine saturated reservoir formation is necessary to understand the long-term fate of injected CO_2 through different trapping mechanisms: Physical trapping, dissolution trapping and mineral trapping. In this work, we investigate CO_2 sorption in brine saturated and dry Mt. Simon sandstone samples via in-situ Fourier Transform infrared spectroscopy (FT-IR). The CO_2 sorption was characterized at elevated pressure from 50 psig to 1200 psig at a temperature of 50ºC. A single symmetrical absorption band, centered at 2342 cm^{-1}, was observed for CO_2 sorption in a brine saturated sample, which is also detected when CO_2 dissolves in the bulk water/brine. However, 2 asymmetric peaks, located near 2330 cm^{-1} and 2360 cm^{-1}, were observed for CO_2 adsorption onto the dry sample. The different position of the detected CO_2 peaks in these 2 cases implies that when CO_2 sorption occur in a brine saturated sample, almost all the CO_2 molecules are dissolved in the brine and that no CO_2 adsorption on the surface is detectable via the FT-IR technique. The amount of CO_2 sorption in the brine saturated sample was quantified by a correlation of the integrated peak area to the CO_2 solubility in the same bulk brine, as measured separately via a PVT-cell approach. Finally, the CO_2 mass transfer in the brine saturated sample, as detected by the FT-IR, was subsequently delineated by a simple mathematical model formulated based on Fick's law for diffusive mass transfer. The presented observations and analysis suggest that the adsorbed CO_2 on the surface of Mt. Simon rocks samples is reduced substantially by the presence of brine in the pore spaces, and that the CO_2 mass transfer in this system is consistent with the molecular diffusion in the brine inside the pore spaces.

566647 Thermodynamic Phase Behavior of CO_2 Storage in Tight Formation

Thermodynamic Phase Behavior of CO_2 Storage in Tight Formation

Kaiqiang Zhang, Lirong Liu, Na Jia

Keywords: CO_2 storage; Thermodynamic phase behaviour; Fractured nanopores; Adsorptions; Analytical equation of state.

Carbon capture and storage (CCS) is currently a main way to reduce anthropogenic CO_2 emissions. However, the supplies of ideal reservoirs for the geological CO_2 storages are substantially less than the demands if the CO_2 storage is treated as a usual method to reduce the anthropogenic CO_2 emissions and implemented worldwide. On the other hand, the tight/shale formations with micro/nanopores are extensively distributed, which are usually fractured by providing the outer pressures (e.g., injecting CO_2) or after the tight/shale oil/gas productions. Hence, it is necessary to study the potential CO_2 storages in the fractured tight formations. In this paper, thermodynamic phase behaviour for the CO_2 storage process in the fractured tight formations (i.e., nanoproes) with adsorptions are studied.

More specifically, first, an analytical equation of state (EOS) is modified and applied to calculate the thermodynamic phase behaviour of confined pure/mixing fluids in nanopores considering the effects of pore radius and molecule–molecule (m–m) interactions. Second, a new empirical correlation for predicting the adsorption thickness in nanopores is initially developed. The modified EOS coupled with the new adsorption thickness correlation and the fracture geometry equation is used to calculate the phase behaviour of confined pure and mixing CO_2 streams in fractured nanopores with adsorptions. The phase behaviour, which include the pressure–volume (P–V) diagrams, pressure–temperature (P–T) diagrams, and critical properties, of 12 pure substances of CO_2, N_2, and alkanes of C_{1-10} and 12 CO_2-dominated binary and ternary mixtures are specifically studied.

A new empirical correlation for the adsorption thickness is developed, which is coupled with a modified van der Waals EOS (vdW-EOS) and the fracture geometry equation to calculate the phase behaviour of confined pure and mixing CO_2 streams in fractured nanopores with adsorptions. The calculated pressures for all cases in nanopores with and without the adsorptions and fractures are reduced with the system volume increases while they are increased by increasing the system temperature with constant compositions, which are always larger than those in bulk phase at small volumes but in good agreement at large volumes. In comparison with the N_2 or C_{1-10}, the pure CO_2 is more easily transited to be a liquid or supercritical phase. Moreover, any additions of N_2 or C_{1-10} into the pure CO_2 increase the vapour/bubble pressures to different extent, especially in nanopores. In the P–V diagrams of pure and mixing CO_2 streams, the pressures are more sensitive to the volume change in pure gas phase but the liquid–gas or liquid–liquid phase are inert to the volume expansions. The pressures in the P–T diagrams are monotonically increased with the temperature in bulk phase and nanopores. The pressure-transition pressures of the C_{1-10} become more sensitive to the temperature with the carbon number increase. In comparison with the N_2 or C_{1-10}, the pure CO_2 is more easily transited to be a liquid or supercritical phase. Any additions of N_2 or C_{1-10} into the pure CO_2 lead the system pressures to increase more with the temperature to different extent, especially in nanopores. Also, the pressures in nanopores become even higher than those in bulk phase for the CO_2-dominated binary and/or ternary mixtures. The critical properties with the adsorptions and fractures are decreased with the pore radius reduction to some certain level, which agree well with the results without the adsorptions or fractures, and reversed to increase afterwards. The bottom pore radius for the coincidences of the critical properties with and without adsorptions and fractures vary for different substances. The critical properties, in comparison with the pure

CO_2 case, are substantially decreased by adding N_2 or C_{1-10}, wherein the N_2 exerts the strongest effect while the effects of the alkanes are weakened with the carbon number increase. Furthermore, the critical properties of the binary and ternary mixtures with and without adsorptions and fractures are similar at r_p = 0.4–1000 nm. Finally, for CO_2 storage projects in the deep tight formations, three optimum strategies are obtained from this study and listed as follows: lager pore radii (by natural or manmade fractures), purified CO_2 streams, and low temperatures.

At the current stage that the experimental methods are incapable of fully exploring the nanoscale phase behaviour, this study is critically important because the CO_2 storage is the key part to mitigate the greenhouse gas emissions and control the climate change to a manage and meanwhile, its applications in the tight/shale formations with micro/nanopores are inevitable in the near future.

575992 Regrid: A Python Toolbox for Reservoir Simulation Reproducibility, Comparison, and Multiphysics Code Interoperability

Geologic storage of CO2 remains a viable means by which global carbon emissions can be reduced at scales necessary for substantive impact. Geologic carbon storage projects exhibit varying concerns (risks) depending on lithology, simultaneous or subsequent utilization, and volumes injected. In most cases, multiphase flow simulations represent a critical management tool to forecast future behavior and also to evaluate reservoir properties.

However, reservoir models are often treated as "black box tools," and modeling often involves as much art as science. As such it can be difficult to quantify uncertainty related to reservoir model simulations. Additionally, often coupled processes are of interest within a carbon storage project which cannot be readily modeled natively within a multiphase simulator. For these reasons, tools that allow for easier code comparisons, reproducibility between investigators, and incorporation of related simulations for coupled process modeling are extremely useful.

The Regrid project is an open source python module with these aims. Regrid allows for mesh conversion between popular simulators including Petrel/ECLIPSE, STOMP, SUTRA, MODFLOW, and TOUGH(2). Additionally, Regrid allows for models to be converted to VTK formats which can serve either as an intermediary format for multiphysics simulations as well as for post-processing and visualization. We present the use of Regrid with application at an active field site under investigation by the National Energy Technology Laboratory and its Southwest Regional Partnership on Carbon Sequestration. As an illustration, we demonstrate linking of reservoir models developed in Petrel with shallow USDW MODFLOW models. Additionally, injection scenarios are compared across simulators. Finally, multiphysics use cases will be shown alongside post-processing visualization.

574863 Field Testing of Emerging Techniques for Active Management and Monitoring of Commercial CO_2 Storage

The Energy & Environmental Research Center (EERC) is partnering with the U.S. Department of Energy (DOE) National Energy Technology Laboratory and project partners to advance techniques that could benefit the active management and monitoring of commercial CO_2 enhanced oil recovery (EOR) and geologic CO_2 storage projects.

Because of a host of technical, social, regulatory, and economic factors, techniques that result in more compact and efficient CO_2 storage in geologic formations are desirable. Results of several modeling studies indicate that active reservoir management (ARM) using brine extraction can reduce stress on sealing formations, geo-steer injected fluids and pressure plumes, divert pressure or fluids from potential leakage pathways or sensitive areas, reduce area of review, and improve geologic CO_2 storage injectivity, capacity and efficiency. The EERC is validating these concepts through a well-designed field test being implemented at an active saltwater disposal (SWD) site that serves as a proxy for geologic CO_2 storage. Simultaneous injection and extraction of brine using the existing SWD wells and a newly installed brine extraction well, combined with a monitoring program and reservoir simulation, are being used to evaluate the ARM scenarios being tested. As a proxy to commercial-scale geologic CO_2 storage, the SWD site injects the equivalent of a quarter million tons of CO_2/year on a volumetric basis which provides an established fluid and pressure plume that ARM performance can be validated against.

The monitoring program includes a time lapse borehole-to-surface electromagnetic (BSEM) survey, which is being conducted as a means of imaging changes in resistivity within the test formation. Brine injected at the site is displaced native saline formation water, creating a resistivity "plume" that can be measured by the BSEM survey. A baseline survey was acquired in the fall of 2018, and a repeat survey will be collected at the end of the field test. The baseline survey is being used to characterize site geology and identify the current distribution of the resistivity plume. The repeat survey will be used to validate reservoir simulations of ARM performance by measuring changes to the resistivity plume as a result of the field test.

The scalable, automated, semipermanent seismic array (SASSA) project led and managed by the EERC was a 3-year proof-of-concept study to evaluate and demonstrate an innovative application of the seismic method. By using a sparse surface array of nodal seismic sensors paired with a single, remotely operated active seismic source at a fixed location, strategically chosen reservoir reflection points were monitored for time-lapse changes that indicated CO_2 saturation changes in the reservoir at those locations, thus monitoring the movement of the saturation front. The application differs from the normal paradigm of collecting a spatially dense data set to produce an image. Instead, standard time-lapse processing and innovative displays of incremental monitor trace data for individual receiver locations were analyzed for signal character changes that could be attributed to the passing of a CO_2 saturation front or, possibly, changes in reservoir pressure. Monitoring was done in a low-impact, cost-effective manner, remotely, with the future intention of automating as many of the processes as possible.

Data collection occurred over the course of 1 year at an oil field undergoing CO_2 injection for EOR and focused on four overlapping "five-spot" EOR injector–producer patterns. Selection, procurement, configuration, installation, and testing of project equipment and collection of five baseline data sets were completed in advance of CO_2 injection within the study area. Weekly remote data collection produced 41 incremental time-lapse records for each of the 96 nodes.

CO_2 injection data and the reservoir simulation results showed that saturation distributions in the study area progressed in a manner that only a subset of the 96 node midpoints could be expected to show character changes due to the presence of CO_2. Data from twenty-six nodes were selected for in-depth analysis. The data sets were affected by several types of seismic noise which presented processing and interpretation challenges. Weather and the cultural noise associated with an active oil field were the most challenging because of their variability.

Interpretation results were encouraging, but mixed. Several nodes showed seismic reflection character changes indicative of the presence of CO_2, while other nodes with corresponding reflection points where no CO_2 was expected showed no effect. However, several nodes showed ambiguous interpretation because of poor signal-to-noise ratios. Validation methods from reservoir simulations and a time-lapse 2-D line acquired through the middle of the study area helped the interpretation but did not fully remove ambiguity.

A second iteration of the project is now under way in a different part of the field and directly addresses over twenty lessons learned during the proof-of-concept study. In particular, two powerful surface orbital vibrator sources that output 10 tons of force each are addressing the signal-to-noise issue. Two fixed source locations provide double the number of subsurface monitored points, and point locations have been chosen in conjunction with the field operator's plans to minimize nodes that will not encounter CO_2. The ultimate aim of the project is to evaluate whether SASSA technology can provide cost-effective monitoring of future CO_2 injection projects.

The EERC has also explored the use of interferometric synthetic aperture radar (InSAR) as a monitor tool for CO_2 EOR. SAR is an airborne or spaceborne radar system that first transmits microwave signals and then receives back the signals that are reflected, or backscattered, from Earth's surface. InSAR is a technique for using high-resolution SAR images to generate high-quality digital terrain elevation maps with phase interferometry methods. Since InSAR can monitor ground deformation, it is possible to assess the impact of injecting CO_2 into the reservoir at Earth's surface.

InSAR data were acquired over a 370-km^2 area that covered an oil field at various phases of EOR development. The Advanced Land Observation Satellite (ALOS) and COSMO-SkyMed (CSK) satellite system were used to acquire data between 2007 and 2011. ALOS acquired 21 sets of InSAR measurements every month from 2007 to 2011 but lost power 2 years before the start of CO_2 injection at Bell Creek. CSK acquired measurements approximately every 16 days starting in 2015. The CKS data, which provide both higher frequency measurements, higher resolution data, longer wavelengths, and lower-noise levels, were used to investigate the applicability of INSaR measurements to monitor CO_2 EOR floods in complex (e.g., seasonal snow cover, challenging topography, vegetation changes, etc.)

A processing technique called SqueeSAR was applied to CSK data to overcome the InSAR challenges associated with atmospheric effects and with the topography, agricultural activities, months of snow cover, and areas of pine forest of the Bell Creek Field. Strong correlations between fluid injection/production and surface elevation changes were observed in areas that corresponded to a compartmentalized distribution of geologic heterogeneities. Correlations between net fluid injection/production and small changes in the surface elevation were improved by removing low-frequency trends and applying time lag to the InSAR time series. Estimated reservoir pressure differences from geomechanical modeling and InSAR data inversion validated the observed correlations.

The work indicates that InSAR techniques may provide a low-cost, low-impact means of better understanding CO_2 EOR pattern performance and reservoir heterogeneities in challenging surface environments.

579052 Innovative Fluidized Bed Technologies in Coal to Chemical Processes

The increasing demand of chemicals including olefins and aromatics in China has triggered significant interests in developing coal via syngas and/or methanol to chemical processes. One of the examples is the methanol to olefins process (MTO) developed by the Dalian Institute of Chemical Physics (DICP), Chinese Academy of Sciences, which was successfully commercialized in 2010 in Shenhua Baotou's coal-to-olefins (CTO) plant (with an ethylene and propylene production capacity of 600 kt/a). In this DMTO process, fluidized bed reactor-regenerator configuration, akin to that in the fluid catalytic cracking (FCC) process, has been used. The DMTO reactor is a shallow fluidized bed with an aspect ratio of 0.3, representing a novel practice in industrial fluidized bed design. In this paper, we will discuss the progress of fluidized bed technologies in DICP, focusing on the applications in coal to chemical processes developed in recent years. We first report the fluidized bed technologies used in DMTO process, which include the shallow bed reactor, catalyst circulation rate control, and pre-coking technology. We then introduce the research and development (R&D) activities on fluidized bed reactor development and scale-up for methanol to propylene (DMTP) and methanol to aromatics processes. Beyond that, our strategy of innovative research for coal to chemical processes will also be highlighted.

576964 Methyl Formate Hydrogenation to Higher Hydrocarbons over Mn- and Ti-Oxide Promoted Cobalt Catalysts

Methyl formate hydrogenation to higher hydrocarbons over Mn- and Ti-oxide promoted cobalt catalysts

Authors: Andrew Raub, Motahare Athariboroujeny, Norbert Kruse

Methyl formate, the simplest of the carboxylic esters, has been considered since the early 1990s as a potential platform molecule for C1 chemistry[1]. Recently, we have established that it is possible to produce an array of chain-lengthened paraffins from methyl formate via a Fischer-Tropsch akin process utilizing cobalt-based catalysts at atmospheric pressure conditions. The fact that this molecule contains CO and H_2 in a 1:1 ratio makes it a potentially suitable feedstock for the preparation of hydrocarbon chains. In addition, methyl formate's low toxicity and relatively high boiling point (32 ^{o}C) make the molecule easy to transport and store. With further exploration, methyl formate hydrogenation may also provide a novel route for the synthesis of olefins and oxygen-functionalized hydrocarbons. In this work, the basic catalytic science of methyl formate hydrogenation over cobalt catalysts promoted by Mn- and Ti-oxides will be investigated. Manganese and titanium were selected due to their documented promoter effects observed for cobalt-based catalysts studied under conventional Fischer-Tropsch synthesis conditions[2, 3] . This paper will provide kinetic and mechanistic insight into the atmospheric pressure hydrogenation of methyl formate using transient response methods.

Chemical Transient Kinetics (CTK), as applied in the present studies, relies upon the abrupt switching between a non-reactive (H_2 + Ar) and a reactive gas mixture (H_2 + He + methyl formate) while maintaining constant molecular flow conditions at the inlet to a catalyst bed. The initial construction of the catalytically active surface, termed the "build-up", occurs when reactants are admitted to the gas phase over a freshly reduced catalyst surface. Likewise, the removal of reactants from the gas phase results in a "scavenging" of the surface as residual adsorbed species are hydrogenated and desorb. Observation of the build-up and scavenging behaviors for a catalytic system provides mechanistic insight unattainable from a steady-state kinetic analysis. Integration of calibrated surface flows allows for time-dependent carbon and oxygen coverages on the catalyst surface to be determined at different stages of the reaction.

During the build-up of the methyl formate hydrogenation reaction, using a Co-MnO_x catalyst under atmospheric pressure conditions at 220 ^{o}C, alkanes are observed to form in the sequence of their carbon numbers. With the exception of methane formation, which is the first product observed in the present study, chain-lengthened alkanes appear once methyl formate decomposes to produce CO in the gas phase. The time-correlation of chain lengthening initiation and CO appearance in the gas phase places a strong argument in favor of a CO insertion mechanism being in operation. Differing from earlier studies on methyl formate hydrogenation in our lab, in which only alkanes were observed during methyl formate hydrogenation over Co/MgO catalysts[4], we also observe olefins. Remarkably, the delay time for olefin formation is longer than that for alkanes. Methanol production is also observed, time-delayed relative to the appearance of gaseous CO. The formation of methanol from methyl formate under mild conditions using heterogeneous promoted cobalt catalysts has not been previously reported.

As for the reaction mechanism, we hypothesize that formyl groups (resulting from the dissociation of methyl formate) adsorbed on the catalyst surface undergo decarbonylation to yield CO, which subsequently inserts into the bonds of surface alkoxy or hydroxyl groups. The resulting carboxylate-type species may then be hydrogenated to produce alkanes/alkenes or turned into a (C+1)-alkoxy homologue subject to another CO insertion, and therefore, chain lengthening. Such a scenario would also be able to account for the formation of oxygenated products.

References

1. Lee, J. S.; Kim, J. C.; Kim, Y. G., Methyl Formate as a New Building Block in C_1 Chemistry. *Applied Catalysis* **1990,** *57* (1), 1-30.
2. Dinse, A.; Aigner, M.; Ulbrich, M.; Johnson, G. R.; Bell, A. T., Effects of Mn promotion on the activity and selectivity of Co/SiO2 for Fischer-Tropsch Synthesis. *Journal of Catalysis* **2012,** *288*, 104-114.
3. Morales, F.; de Smit, E.; de Groot, F. M. F.; Visser, T.; Weckhuysen, B. M., Effects of manganese oxide promoter on the CO and H-2 adsorption properties of titania-supported cobalt Fischer-Tropsch catalysts. *Journal of Catalysis* **2007,** *246* (1), 91-99.
4. Iablokov, V.; Kruse, N., Discovery of a Fischer-Tropsch Hybrid Reaction: Hydrogenation of Methylformate to Long-Chain Hydrocarbons with Anderson-Schulz-Flory Chain Length Distribution. *Chemcatchem* **2019,** *11* (4), 1200-1204.

570450 Gallate-Based Metal-Organic Framework As a Heterogeneous Catalyst for the Coupling of CO_2 and Epoxides

Carbon dioxide (CO_2) is the major greenhouse gas and its increasement in the atmosphere has evoked global environmental and societal concerns. The chemical transformation of CO_2 into value-added chemicals has attracted significant interest due to its great potential as a nontoxic, ubiquitous and sustainable C_1 resource. [1] The conversion of CO_2 into cyclic carbonates, which have wide applications as pharmaceutical, fine chemical intermediates, monomers for the production of polycarbonates, and aprotic polar solvents, is one of the most promising strategies because of its high atom efficiency. Although homogeneous catalysts have been industrialized for the production of cyclic carbonates, the processes still suffer from unviable catalyst recovery process and difficult product purification as well as extreme pressure and temperature conditions. [2] Hence, the design and preparation of highly efficient heterogeneous catalyst is of great significance to boost the cyclization of CO_2 and epoxides. Metal-organic frameworks (MOFs), which feature structural versatility, modularity, and amenability to be designed with specific functionality, have received much attention in catalysis, adsorption, and so on. [3-4] In view of their ordered porous structure, hydrophobic confined environment, and fruitful functionalizations, MOFs have been explored as one of the most appropriate materials for heterogeneous catalysis.

In the present work, we employed gallate-based metal-organic frameworks (M-gallate, M= Ni, Co, Mg, Mn) as a highly efficient and recyclable heterogeneous catalyst for CO_2 cycloaddition with epoxides in the presence of quaternary ammonium salts under solvent-free conditions. Experimental investigations revealed that Co-gallate shows superior catalytic activity to other M-gallate (M= Ni, Mg, Mn) and cobalt salts. This protocol is compatible with a plethora of terminal epoxides and consequently provides a practical alternative method for the fixation of industrial CO_2 into value-added cyclic carbonates.

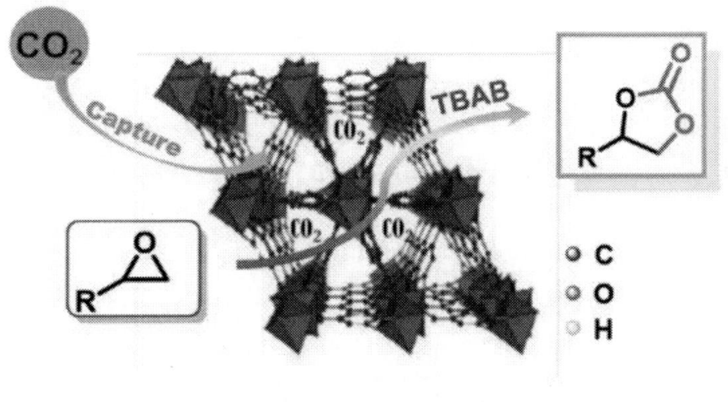

Scheme 1. M-gallate catalyzed CO_2 cycloaddition with epoxides

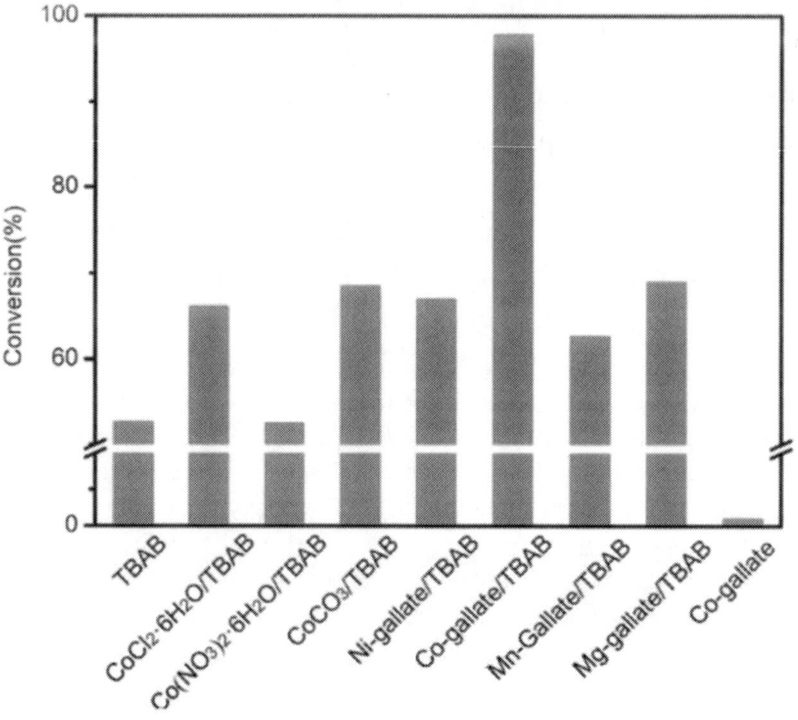

Figure 1. CO_2 cycloaddition with epoxide with different catalysts

References

[1] Song QW, Zhou ZH, He LN. Efficient, Selective and Sustainable Catalysis of Carbon Dioxide. *Green Chem*. **2017**; 19 (16): 3707-3728.

[2] Gao CY, A J, Tian HR, et al. An ultrastable zirconium-phosphonate framework as bifunctional catalyst for highly active CO_2 chemical transformation. *Chem. Commun*. **2017**, 53, 1293-1296.

[3] Kurisingal JF, Babu R, Kim S-H, et al. Microwave-induced synthesis of a bimetallic charge-transfer metal organic framework: a promising host for the chemical fixation of CO_2. *Catal. Sci. Technol*. **2018**; 8, 591-600.

[4] Babu R, Roshan, R, Kathalikkattil AC, et al. Rapid, Microwave-Assisted Synthesis of Cubic, Three-Dimensional, Highly Porous MOF-205 for Room Temperature CO_2 Fixation via Cyclic Carbonate Synthesis. *ACS Appl. Mater. Interfaces*. **2016**; 8, 33723-33731.

557675 Biological Upgrading of Methane for the Production of Chemicals and the Techno-Economic Analysis

The current crisis of global warming is primarily attributed to CO2 production from excessive use of fossil fuels during recent decades, and has increased demand for renewable biofuels tremendously. Lipids are drawing considerable attention in relation to the production potential of biodiesel on the basis of their nontoxic, sustainable, and energy efficient proprieties. However, the high cost of microbial lipid produced by oleaginous microorganisms mainly stems from the high cost of glucose, which is estimated to be about 80% of the total medium cost. Therefore, considerable efforts have been directed toward minimizing the carbon source cost and finding new alternative carbon sources. In this report, methane derived from biogas or natural gas was investigated for chemical production. After developing the culture modes and optimizing the culture conditions, both high titer and productivity were achieved in high cell density cultures of different microorganisms using various carbon sources.

In light of the relatively low price and the tremendous amount of methane and increasing demands of biochemicals and liquid transportation fuels, attention has begun to turn to novel biocatalyst for conversion of methane into bio-based products . A techno-economic analysis (TEA) was performed for an integrated process using fermentation process converting methane into chemicals such as lactic acid and isobutanol, which can be used as the pursuer for the production of chemicals and fuels. The entire process has been projected by using the Aspen plus software and the costs were calculated using modified excel spreadsheet. A sensitivity analysis was carried out based upon the identified key parameters, such as plant capacity, process technology (yields and rates), productivity, and raw material cost. This presentation also includes single-point sensitivity analyses and scenarios analyse results for several prospective cases. The presentation will also reveal that the major driving forces from operating cost and capital cost. The efforts of reducing the greenhouse emission from this process was estimated and compared with the traditional combustion of biogas and natural gas.

563733 In-Situ Hydrogen Sulfide Removal and Biogas Upgrading Via a One-Reactor System Integrating Solid-State Anaerobic Digestion of Food Waste and Bioelectrochemical Treatment

Food waste (FW) is estimated to account for 40% of the food produced in the United States, making it a critical and urgent environmental concern with a magnitude of 280-300 kg per year per capita. Among the approaches to valorize FW for biochemical/bioenergy production and to reduce its environmental footprint, solid-state anaerobic digestion (SSAD, total solids content = 15 wt%) offers the superiorities such as smaller reactors required, lower energy input for heating and stirring, easier disposal of digestate with lower moisture as biofertilizer or soil amendment, etc. However, with the higher organic contents accompany some shortcomings, typically including easy digester acidification and even failure caused by accumulation of excessive volatile fatty acids (VFAs), buildup of hydrogen sulfide (H_2S) in the biogas that is toxic to personnel and corrosive to facilities, etc. Conventional countermeasures such as extra addition of water or some bulk agents (e.g., yard wastes) reduce the capacity of FW treatment.

To fill in this knowledge gap, an innovative one-reactor system incorporating SSAD of FW and bioelectrochemical (BEC) treatment of leachate was studied. In the upper layer of this reactor, FW was hydrolyzed and VFAs were produced and washed down into the leachate. In the lower layer of reactor inserts the BEC unit. A pair of electrodes made of low carbon steel were submersed into the leachate and applied a voltage of 0.7 V, which secured the removal of most H_2S with relatively less material consumption of electrodes and increased the buffering capacity of leachate. A cyclic flow of leachate was created so that the BEC-treated leachate with a stronger buffering capacity was pumped back to upper layer for a more thorough mass transfer of SSAD. Moreover, the biogas profile was upgraded since the BEC treatment contributed to the reduction of some CO_2 to CH_4. Consequently, in-situ H_2S removal, biogas upgrading, and process stabilization were simultaneously achieved through this one-reactor design integrating SSAD of FW and BEC treatment of leachate.

569653 Copper Speciation in Cu-Mordenite Catalysts for the Oxidation of Methane to Methanol

Methane from abundant natural gas reserves represents a valuable potential feedstock for the chemical industry, but the requirement of liquefying natural gas for storage and transport renders much of the gas present in remote deposits economically inaccessible. Copper-exchanged zeolite catalysts are capable of chemically converting CH_4 to higher-value liquid CH_3OH with high selectivity, but low yields have prevented industrial application and hampered active site characterization. Utilizing the H^+ form of the zeolite instead of the Na^+ form has been shown to increase CH_3OH productivity, which should facilitate identification of the Cu active site, but the active site structure in Cu-H-MOR has not yet been unambiguously defined. Under conditions of high CH_4 pressure, the C1 product yield during cyclic operation of the Cu-H-MOR catalyst was 0.42 mol (mol Cu)$^{-1}$. Linear combination fitting of the Cu K-edge X-ray absorption spectrum of the Cu-H-MOR catalyst showed that 83% of the Cu was auto-reduced in He, which is assumed to be the redox-active fraction of Cu for CH_3OH formation. Normalizing the product yield to the redox-active fraction of Cu gave a reaction stoichiometry of 0.50 mol (mol Cu)$^{-1}$, consistent with the presence of dicopper active sites. The O_2-activated dicopper sites were further characterized by EXAFS, UV-vis and resonance-Raman spectroscopy.

577349 Dual Production of Higher Hydrocarbons and Syngas from Greenhouse Gases in a Hydrogen-Permeable Membrane Reactor

CH_4 and CO_2 are the largest anthropogenic greenhouse gas (GHG) contributors to climate change. Due to the abundance of natural gas, CH_4 will be increasingly utilized for both energy and chemical production with potentially even greater GHG emissions unless methods are developed to increase the efficiency of CH_4 conversion, ideally by non-oxidative (no CO_2 production) processes. In this talk, we present the first ever direct non-oxidative methane conversion (DNMC) with CO_2 to produce value added C_{2+} hydrocarbons in an integrated DNMC-catalyst/hydrogen-permeable-membrane reactor. Specifically, a direct utilization of CH_4 and CO_2 to simultaneously produce C_{2+} hydrocarbons (C_2 and aromatics) and syngas (CO and H_2) on opposite sides of a mixed ionic-electronic conducting $SrCe_{0.7}Zr_{0.2}Eu_{0.1}O_{3-d}$ membrane reactor will be discussed. On one side (interior) of the membrane reactor, direct non-oxidative methane conversion (DNMC) over an iron/silica catalyst produces C_{2+} hydrocarbons and H_2. On the other side (outer surface) of the membrane, permeated hydrogen (driving the DNMC reaction) reacts with a CO_2 sweep gas to form CO and water via the reverse water gas shift (RWGS) reaction. This novel single hydrogen-permeable membrane reactor simultaneously addresses both reduction of greenhouse gas (CO_2 and CH_4) emissions as well as production of value-added hydrocarbon products (C_{2+}, CO, and H_2) with *in situ* gas separation. This membrane reactor not only circumvents gas-to-liquid (GTL) thermodynamic limitation of CH_4 conversion to C_{2+} hydrocarbons (without producing byproduct CO_2) but in fact uses CO_2 as a sweep gas to react with the permeated hydrogen to form CO as a feedstock for additional GTL reactions, thus simultaneous producing value added chemicals while reducing GHG emissions.

574681 Elucidating the interactions of H_2S and CO_2 with transition metal sulfide catalysts for natural gas upgrading

Natural gas reserves contain large concentrations (>10 % mol) of sour gas molecules, H_2S and CO_2. These streams are treated using energy intensive separation methods to remove sour gas molecules and to preserve downstream catalysts and equipment. We investigate sour gas tolerant catalysts with emphasis on metal sulfides to devise catalytic processes for sour natural gas monetization without the need for CO_2 and H_2S removal. Metal sulfides have been shown to be active for methane coupling, alcohol production and alkane dehydrogenation in addition to its traditional use in hydrotreating. Specifically, MoS_2 is used in this work to investigate the reverse water gas shift (rWGS) chemistry with the goal of activating CO_2 molecule in the presence of H_2S. In particular, we will present fixed bed reactor studies and transient kinetics experiment with and without H_2S cofeed to measure apparent barriers and reaction orders. Further, in to complement with the experimental data, we will also present detailed *ab initio* phase diagrams of MoS_2 edges. Next, we will discuss the effect of acid gases (CO_2, H_2S) on reactivity-selectivity of these transition metal sulfide catalyst for natural gas upgradation. Our results suggest that the presence of H_2S induces change in catalyst structure (active sites) and furthermore, it influences the activity and selectivity towards the reaction.The insights from this work can, therefore, will be used to evaluate a spectrum of transition metal sulfide catalysts for upgrading C_1-C_4 hydrocarbons in the presence of CO_2 and H_2S.

Acknowledgment:

This work was supported as part of UNCAGE-ME, an Energy Frontier Research Center funded by the U.S. Department of Energy, Office of Science, Basic Energy Sciences under Award No. DE-SC0012577.

574954 Understanding the Synergistic Effects of Feed Composition and Metal Oxide Formulation in a Chemical Looping Oxidative Coupling of Methane System

The abundance of methane in the form of shale gas greatly incentivizes the direct conversion of methane to value added products such as olefins. Chemical looping oxidative coupling of methane (CL-OCM) is a promising pathway for converting methane to hydrocarbons such as ethylene. The novelty of the chemical looping mode lies in methane being activated by a metal oxide, instead of molecular oxygen as done in the conventional co-feed OCM to produce ethylene as the desired product. The metal oxide thus serves as a catalytic oxygen carrier (COC), which provides lattice oxygen to oxidize hydrogen and dopant-induced oxygen vacancies serve as active sites for methyl radical formation and desorption. Previous studies have highlighted the mechanism of CL-OCM and the key reactions responsible for the production of ethylene on Mg_6MnO_8 COC. This information aided in a rational modification of the base COC resulting in the overall hydrocarbon yield of 28.6% for a modified Mg_6MnO_8 COC.

The objective of this study is to further optimize the modified Mg_6MnO_8 COC and investigate the synergistic effect of different feed compositions on the performance of CL-OCM by a combined computational-experimental approach. The effect of different supports for the COC on methane conversion and hydrocarbon selectivity has been investigated experimentally in a fixed bed reactor. These results are corroborated with solid characterization tools such as X-ray diffraction (XRD), X-ray photoelectron spectroscopy (XPS) to look at the bulk and surface changes. Thermogravimetric analysis (TGA) helps in characterizing the kinetics of lattice oxygen consumption on different supports. *Ab initio* atomistic simulations and density functional theory calculations were carried out to explore to the role of supports in enhancing the activity of COC. Further, the effect of co-feeding ethane with methane has also been investigated on the supported and un-supported COC. The trade-off between over-oxidation to carbon oxides versus OCM reaction temperature due to ethane in the feed has also been explored. This study provides a comprehensive understanding of the CL-OCM system and enables further development of this technology.

568984 Mo Oxide Supported on Sulfated Hafnia: Novel Catalyst for Activation of Propane

Mo oxide supported on sulfated hafnia: novel catalyst for activation of propane

Ashraf Abedin[1], Swarom Kanitkar[1], Srikar Bhattar[1], James J. Spivey*[1]

[1]Cain Department of Chemical Engineering, Louisiana State University, Baton Rouge, LA, USA

For the past few decades the production of natural gas in the U.S. has increased dramatically after the introduction of modern innovations such as hydraulic fracturing, offshore drilling, and horizontal drilling. Propane, the third largest constituent of natural gas, has caught significant attention for it's conversion towards value added chemicals. Much of the effort has been devoted to indirect conversion approaches like oxidative dehydrogenation and partial oxidation catalytic processes. However, there has been much interest in direct conversion of propane to higher hydrocarbons, by getting rid of any intermediate steps that require additional energy supply.

Propane aromatization is a direct approach to activate propane and produce higher hydrocarbons like propylene, benzene etc. Pt or Ga based HZSM-5 catalysts have shown prominent results in literature for propane aromatization. These are known as bifunctional catalysts, where the metal sites in their reduced states activate propane to produce alkyl intermediates, which subsequently can get oligomerized on the Brønsted acid sites provided by HZSM-5. In this work Mo is used as the active metal site supported on sulfated hafnia (SH) - a novel acidic support for this reaction system. Mo oxide is reduced in situ to form Mo oxycarbides/carbides phase to activate propane. Sulfated hafnia is homologous to sulfated zirconia, which is a highly acidic support for alkane activation. Even though the shape selectivity feature of HZSM-5 is not present in sulfated hafnia, the acidity is believed to be enough to generate higher hydrocarbons from the alkyl intermediates provided by the Mo sites.

5% Mo-SH was synthesized via impregnation method before it was characterized using DRIFTS and ammonia TPD to study the catalytic acidity after Mo loading. The nature of the active Mo sites was studied by L_{III} edge XANES and HR-TEM. Prior to initiating aromatization reaction, the catalyst was heated up to the reaction temperature under Argon. The catalyst was reduced under hydrogen for an hour before introducing propane into the reactor system. Reaction was run for ~700 mins and the products were analyzed using GC-MSD system. Primary products observed were ethylene, ethane, propylene, butane and a small amount of benzene. Deactivation occurred primarily due to carbon deposition, confirmed later by subsequent TPO analysis on spent catalyst.

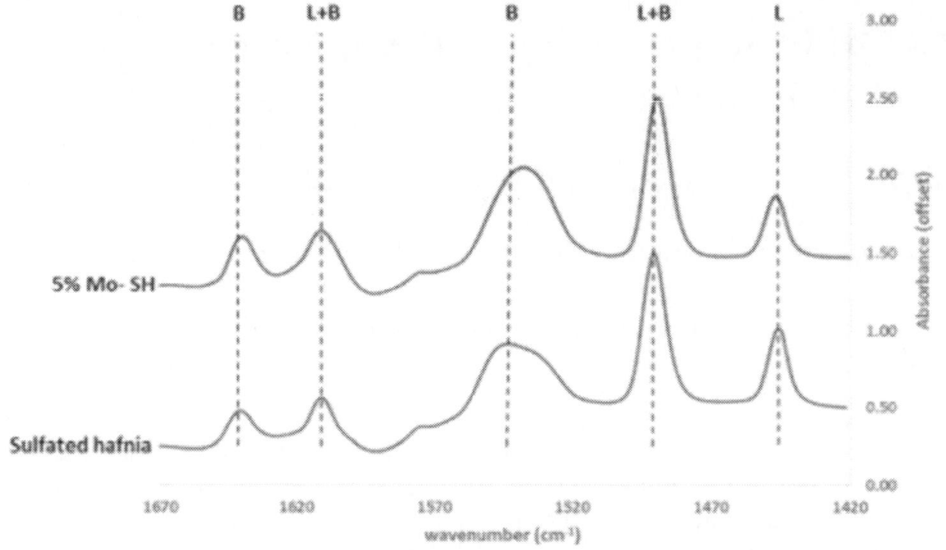

Figure 1: Pyridine-DRIFTS comparison data for sulfated hafnia and Mo-SH catalysts. Samples were desorbed in pyridine at 100^0C. (a) SH (b) 5% Mo-SH ; B- BrØnsted, L= Lewis

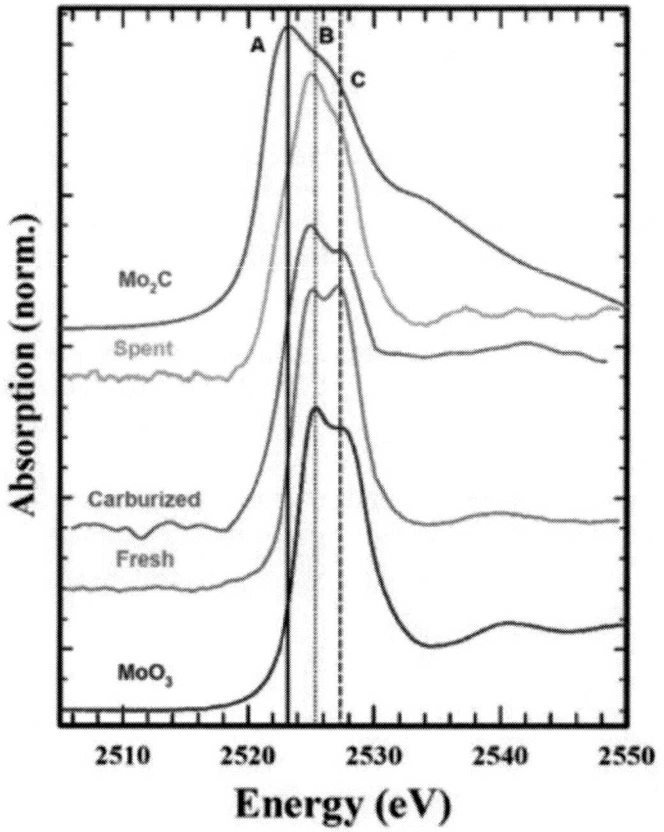

Figure 2: L_{III} edge XANES spectra for catalysts and reference samples: (a) MoO_3 standard, (b) Fresh 5% Mo-SH (oxidized state), (c) Carburized 5% Mo-SH, (d) Spent 5% Mo-SH, (e) Mo_2C standard

570919 Nanographene By Microwave-Assisted Plasma-Mediated Methane Pyrolysis: Spectroscopic and Microscopic Characterization

Microwave (MW) – driven plasma – mediated decomposition of methane is demonstrated here to produce different signature forms of carbon. Focused MW energy creates an intense plasma zone reaction driving methane decomposition to produce clean H_2, not requiring purification or downstream concentration stages. Moreover MW-driven plasma processing does not produce CO_2 nor require water, in contrast to steam methane. That solid carbon is produced not only accomplishes carbon capture, but recent results have shown the production of premium carbons, including graphene [1–5]. Other forms of carbon include graphitized and amorphous carbon particles. This work is the first to demonstrate control of the product's morphology by controlling the ratio of reactants relative to other gas-phase processes of combustion [6,7] or pyrolysis [8,9]. Contrary to combustion and pyrolysis processes where a higher H/C ratio typically decreases the formation of aerosolized carbon (soot), a higher H/C ratio in the reactant gas mixture is shown to net a higher relative yield of graphitized product, particularly graphene.

The current work uses a MW-driven plasma-mediated environment with a gas feed comprising of argon, methane, and hydrogen in varying proportions to study the effect of hydrogen concentration on the yield and quality of graphene sheets so formed. Graphene sheets produced appear relatively small in size with an X-Y dimension of 50-100 nm, but are interconnected, resembling a form of "crumpled" graphene [4,5], though the contributions of material collection and TEM sample preparation to this structure are not well known presently. For reason of the apparent lateral dimension of the sheets, it is hereafter referred to as nano-graphene (NG). Notably the plasma process also leads to the formation of amorphous and semi-graphitic carbon particles with the relative yield of these three carbon "phases" – NG, amorphous and semi graphitic particles – changing in response to the reactor conditions. All three phases have been characterized for their relative contribution to the overall carbon yield while the NG is further studied for its quality that also evolves with changing process conditions. Changes in the carbon product's phase quantities (i.e. relative yields) and phase purity (i.e. quality) are assessed using multiple independent analytical techniques. With focus upon the NG product microscopy and image analysis, Raman, infra-red (FTIR) and electron energy loss spectroscopy (EELS) along with X-ray diffraction (XRD) and thermal analysis show measurement consistency for yield. Phase-pure forms of carbon differing only in their sp^2 edge to basal sites content are introduced as calibration standards for thermogravimetry, thus developing this well-known technique as a tool for quantification of sp^2 carbon phases, to the authors' best knowledge, a first demonstration of such.

Pristine graphene has superior properties of electrical, thermal properties, and mechanical strength. Realization of near-pristine graphene without oxides left over as byproducts of the graphene oxide form and only partially restored sp^2 aromaticity. In contrast to the well documented liquid-based processes, gas phase aerosol production requires no harsh chemicals or secondary processes to recover the sp^2 framework. Applications require material property characterization, particularly for new forms. Additionally, production scale up requires mapping process parameters to the product properties, in this case reactant H_2 and CH_4 concentrations to graphene physical-chemical characteristics.

H Quest's microwave plasma pyrolysis process presents a transformational solution to the challenge of efficient, clean, and cost-effective methane decomposition. Unlike other approaches, it does not rely on conventional (contact, convective, or dielectric) heating or use of thermal plasmas. Rather, it employs a microwave resonant cavity to create localized and high-energy reaction zones in the gas as it passes through the reactor's active zone. Microwaves enable volumetric, non-contact energy transfer to the reactant flow, which is not achievable with radiative or conductive heating in furnaces, by accelerating free electrons in the partially ionized low-temperature plasma. Through electron-molecule collisions, these electrons both transfer the microwave energy to methane molecules and help overcome the high activation energy required by the rate limiting step of hydrocarbon (methane) pyrolysis – the endothermic cleavage of C-C and/or C-H bonds. This results in rapid, direct conversion of methane to chemically active species under atmospheric pressure and mild bulk temperatures: at least 500 lower than conventional decomposition methods.

Additional microwave plasma process advantages include: rapid startup and shutdown, enabling use of intermittent renewable power sources; inherent modularity, scalability, and reduction of costs and risks through replication of a single high-throughput low-cost modular unit reactor; a higher level of safety thanks to lower temperatures and ambient pressures; and simplicity of power delivery [10]. With its prototype microwave plasma reactor, H Quest has demonstrated direct conversion of methane to hydrogen, higher-value chemicals (including acetylene and ethylene), and carbon products: conductive carbon black and a wide range of carbon/graphitic morphologies and nanostructures, including graphene.

Demonstrated production of a high-value carbon, nanographene from natural gas is the key, unprecedented breakthrough achieved by this microwave plasma process. While the graphene market is projected to continue to grow rapidly afterwards, approaching $2 billion by 2025, the extraordinarily high cost of conventional methods of production of graphene material is single greatest factor inhibiting the growth of the markets for these otherwise highly desired and versatile materials. With the 1-2 order of magnitude reduction in production costs we expect market growth to accelerate and new markets to open in industries where the use graphene has hitherto been cost-prohibitive. At present, structural applications are considered to be the largest growth area for this material, but given reports of e.g. five-fold increases in Li-ion battery capacity enabled by this material [11], energy storage may well be the next major market.

H Quest Vanguard is developing broad-spectrum microwave plasma processes targeting conversion of hydrocarbon feedstocks such as coal and natural gas to value-added materials, chemicals and fuels [12,13]. A microwave pyrolysis approach was originally proposed in response to the DARPA initiative for green-house gas (GHG)-emission-free and cost-effective production of US Air Force jet fuel from the domestic coal resources. In H Quest Vanguard's process, natural gas can be used in single-stage reactor as a hydrogen source, eliminating external hydrogen production units and the associated CO_2 production, water consumption, and capital costs, and providing excess hydrogen sufficient for downstream hydro-treating.

Acknowledgements

H Quest Vanguard, Inc. is a privately held technology company, based in Pittsburgh, Pennsylvania, focused on the development and commercialization of novel hydrocarbon conversion technologies. This material is based on work supported by the Department of Energy, Office of Science through sub-award agreement no. 212392 with H Quest Vanguard, Inc. under the Prime Award DE-SC0018703 Phase I STTR, and through sub-award agreement no. 209342 with H Quest Vanguard, Inc. under the Prime Award DE-SC0017227 Phase II SBIR.

References

1. B. Fidalgo, Y. Fernández, A. Domínguez, J. J. Pis, and J. A. Menéndez. Microwave-assisted pyrolysis of CH4/N2 mixtures over activated carbon. *J. Anal. Appl. Pyrolysis* **82,** 158–162 (2008).
2. J. J. Strohm et al. Heavy Fossil Hydrocarbon Conversion And Upgrading Using Radio-Frequency or Microwave Energy. (2012).
3. G. L. Skoptsov, J. J. Strohm, and B. Q. Roberts. Wave modes for the microwave induced conversion of coal. (2015).
4. A. Dato, V. Radmilovic, Z. Lee, J. Phillips, and M. Frenklach. Substrate-Free Gas-Phase Synthesis of Graphene Sheets. *Nano Lett.* **8,** 2012–2016 (2008).
5. A. Münzer, L. Xiao, Y. H. Sehlleier, C. Schulz, and H. Wiggers. All gas-phase synthesis of graphene: Characterization and its utilization for silicon-based lithium-ion batteries. *Electrochim. Acta* **272,** 52–59 (2018).
6. H. Wang and M. Frenklach. A detailed kinetic modeling study of aromatics formation in laminar premixed acetylene and ethylene flames. *Combust. Flame* (1997). doi:10.1016/S0010-2180(97)00068-0
7. C. S. McEnally, L. D. Pfefferle, B. Atakan, and K. Kohse-Höinghaus. Studies of aromatic hydrocarbon formation mechanisms in flames: Progress towards closing the fuel gap. *Prog. Energy*

Combust. Sci. **32,** 247–294 (2006).
8. J. C. Bokros. The structure of pyrolytic carbon deposited in a fluidized bed. *Carbon N. Y.* **3,** 17–29 (1965).
9. J. C. Bokros. Deposition, Structure, and Properties of Pyrolytic Carbon. in *Chemistry and Physics of Carbon-A Series of Advances* (ed. Walker, P. L.) 1–110 (Marcel Decker Inc., 1969).
10. Dagle, R. A., Dagle, V., Bearden, M. D., Holladay, J. D., Krause, T. R., Ahmed, S. (2017). *An Overview of Natural Gas Conversion Technologies for Co-Production of Hydrogen and Value-Added Solid Carbon Products* (No. PNNL-26726; ANL-17/11). Pacific Northwest National Lab.(PNNL), Richland, WA (United States); Argonne National Lab, (ANL). Argonne, IL (United States).
11. Son, I. H., Park, J. H., Park, S., Park, K., Han, S., Shin, J., Choi, J. W. (2017). Graphene balls for lithium rechargeable batteries with fast charging and high volumetric energy densities. *Nature Communications*, *8*(1), 1561.
12. Skoptsov, George L., Alan A. Johnson. *Method for processing hydrocarbon fuels using microwave energy*. U.S. Patent 9,682,359, issued June 20, 2017.
13. Strohm, J.J., Linehan, J.C., Roberts, B.Q., McMakin, D.L., Sheen, D.M., Griffin, J.W. Franz, J.A., Battelle Memorial Institute, 2012. *Heavy Fossil Hydrocarbon Conversion and Upgrading Using Radio-Frequency or Microwave Energy*. U.S. Patent Application 13/401,216.

572306 Low-Thermal-Budget Conversion of Natural Gas Integrated with Industrial Electrification

The shale gas revolution in the United States in recent years has resulted in an oversupply of methane and ethane, the major components of natural gas and natural gas liquids, respectively, and this trend will continue for the foreseeable future. Due largely to these cheap resources, the chemical industry has moved manufacturing plants back to the United States, creating thousands of jobs, while enjoying healthy margins compared to international rivals that typically use more expensive oil-based naphtha as feedstock. However, the ethylene production from ethane or naphtha is energy intensive and represents the single most energy consuming process in the chemical industry. To achieve significant progress in the reduction of both the energy consumption and the carbon footprint for ethylene production, especially from abundant natural gas (NG) or natural gas liquids (NGLs) resources, simple process optimization may not be sufficient, owing to the maturity of the manufacturing industry (centralized and vertical integrated), where materials and energy efficiencies have been extensively optimized with a long track record of reliable operation. Therefore, it is vital to develop transformative methods that are capable of fully exploiting the potential of NG/NGLs as a feedstock. Using ethane, one of primary components in NGLs, as the feedstock, we demonstrated that electrochemical activation of ethane for co-production of ethylene and hydrogen at reduced temperatures with significant reduction in process energy and CO2 emission, compared to the commercial ethane steam cracking process. If the heating value of produced hydrogen is taken into account, the electrochemical deprotonation process actually has a net gain in processing energy. This electrochemical activation process could have significant impact when integrated with renewable energies towards industrial electrification.

576548 Transforming Living Methanotrophs into Chemical Reactors

Reducing carbon emissions means addressing and utilizing carbon and energy from waste. Catalytic conversion of waste-derived, low-value carbon sources, e.g., CO_2 and CH_4, into high-value product streams under mild process conditions represents an economically viable route. Biocatalysts, such as live microorganisms, provide an attractive option for small-scale gas upgrading. The economic viability of biocatalytic approaches is evaluated by the volumetric product synthesis rates of the bioreactors, which is expected to be multiple grams of product per liter per hour. However, current aqueous stirred-tank bioreactors do not achieve high volumetric productivity in an energy efficient manner due largely to poor mass transfer from the gas to the liquid phase. To reach the required volumetric productivity a new bioreactor technology must be developed which increases mass transfer rates by 1-2 orders of magnitude relative to stirred vessels.

At LLNL, we have demonstrated efficient methane gas upgrading to methanol by printing enzyme-loaded polymeric materials. This is accomplished by advanced manufacturing techniques, which created unprecedented 3D gas-permeable microarchitectures that increase gas-to-liquid mass transfer by over an order of magnitude relative to current industrial methods.

To move beyond using expensive enzymes, we have developed new materials by entrapping live methanotroph cells in biocompatible, polymeric and 3D-printable materials, and demonstrated that the cells remained viable for a few days. We are using an engineered methanotroph strain (*Methylococcus capsulatus*) to convert methane gas into liquid organic acids, such lactic acid and muconic acid, both of which are of industrial interest. The first hurdle to entrapping whole cells in polymer is determining the material properties and biocompatibility. We will present the dissolved gas permeability of various formulations of porous silicone and polyethylene glycol, as well as the relationship between material thickness and gas transport. Additionally, we will show the biocompatibility of the materials by assessing the viability of entrapped methanotrophs. Finally, we will present data showing that the entrapped live methanotrophs are able to convert methane into the liquid organic acid products of interest.

This work was performed under the auspices of the U.S. Department of Energy by Lawrence Livermore National Laboratory under Contract DE-AC52-07NA27344. (LLNL-ABS-771899)

574303 Techno-Economic Analysis of Direct Methane Aromatization for Natural Gas Upgrading with Mini-Plant Construction for Remote Deployment

Inexpensive sources of natural gas have led to a boom in development of technologies taking advantage of C1 chemistry; however, a significant number of potential natural gas sources are stranded and lack the infrastructure for economically-feasible transportation. The resulting gas, which is normally flared can, in theory, be recovered at the well-site, purified, and then processed to a product which is easier to transport. Gas-to-liquid (GTL) processes exist which can convert the feed to liquid products which can be more easily transported, although there are process constraints due to the need for deployment in remote locations, and several are being commercialized.[1] One such process, direct methane aromatization (DMA), produces significant amounts of benzene, a valuable commodity chemical, and hydrogen. The benzene can be purified on-site and then transported by truck; the hydrogen can be dealt with on-site as well by either transforming it to another transportable chemical, such as ammonia or cyclohexane, or using it to provide heating or electricity.

Previous economic studies of DMA have been reported which either invoke unrealistic assumptions in the process model,[2] do not consider processing of the gaseous hydrogen,[2, 3] or do not consider the limitations imposed on the process due to remote deployment.[4] With these issues in mind, a DMA mini-plant processing 20 MMSCFD of natural gas was designed with currently available technology to provide a platform for the investigation of process intensified alternatives. The reactor section of the plant uses a DMA catalyst developed at Texas Tech which deactivates at a significantly slower rate than previous catalysts.[5] The unit was investigated for economic feasibility and sustainability. In particular, the analysis identifies four particular areas which can benefit from intensification: aromatics yield (ie. one-pass conversion), coke selectivity, hydrogen recovery, and separation and purification of the aromatic products. Although intensified reactor technology for the DMA process is already being investigated for improving the yield with[6, 7] and without integrated hydrogen recovery[8, 9], the process model framework should provide opportunities for investigating additional intensification technologies, particularly in separations.

1. H. Fleisch. [White Paper], World Bank-Global Gas Flaring Reduction Partnership, 2015.
2. Pérez-Uresti, S.I. *et al.*, *Processes*, 2017, 5, 33.
3. Camilo Corredor, E., P. Chitta, and M. D. Deo. *Fuel Pro. Tech.*, 2019, 183, 55 – 61.
4. Huang, K. *et al.*, *Joule*, 2018, 2, 349 – 365.
5. Rahman, M., A. Sridhar, and S.J. Khatib. *Cat. A Gen.*, 2018, 558, 67 – 80.
6. Kee, B. *et al.*, *Eng. Chem. Res.*, 2017, 56, 3551 – 3559.
7. Morejundo, S.H. *et al.*, *Science*, 2016, 353(6299), 563 – 566.
8. Brady, C., B. Murphy, and B. Xu. *ACS Catal.*, 2017, 7(6), 3924 – 3928.
9. Cao, Z. *et al.*, *Chem. Int. Ed.*, 2013, 52, 13794 – 13797.

572855 Non-Oxidative Methane Conversion: Coupling Gas-Phase and Surface Models of Fe Single-Site Catalysts

Introduction

A better utilization of methane gas would lead to increased environmental and economic performance, particularly for resources that are currently flared. Non-oxidative conversion of methane (NOCM) has especially drawn considerable amount of interest for its simplicity and high selectivity. However, coke accumulation has been the biggest problem for NOCM.

A single site catalyst (SSC) may potentially resolve this issue by preventing adsorbate-adsorbate interaction for a coke growth with its sparsely located active sites [1]. In fact, Guo et al. [2] has shown a promising single site Fe catalyst on SiO_2 which showed stable catalytic activity without forming coke. Additionally, contribution from homogenous reactions to NOCM should not be overlooked because of a high NOCM temperature where homogeneous reactions actively occur. Therefore, we have combined SSC chemistry with gas-phase chemistry to model a NOCM process to comprehensively screen conditions for poly aromatic hydrocarbon (PAH) production which is indicative of coke formation.

Materials and Methods

Kinetic parameters of elementary steps for catalytic methane activation on SSC were calculated by *ab initio* density functional theory (DFT). A single Fe center with two neighboring carbons was modeled in a 2×2 cristobalite supercell. A Grid-Based Projector-Augmented Wave (GPAW) method with revised Perdew-Burke-Ernzerhof (RPBE) exchange-correlation functional was used in Atomic Simulation Environment (ASE). Barrier energies were calculated by finding a transition state with constrained minimization method.

For the gas-phase chemistry, the detailed kinetic mechanism was obtained from a recent literature study on methane pyrolysis to PAH formation [3]. Finally after combining surface chemistry species and gas-phase species into a single input file, Cantera was used to simulate constant pressure, isothermal batch reactor [4].

Results and Discussion

A complete free energy diagram with barrier energies for surface methane activation was obtained. The elementary reactions include methane dissociation, methyl radical desorption, and H_2 dissociation. The methane dissociation steps had the highest barriers.

Coupled heterogeneous and homogeneous reaction simulation was carried out with kinetic information calculated from barrier energies. The simulation was done for various reaction conditions including reaction time, pressure, inlet hydrogen concentration, and temperature to broadly illustrate different combinations of the controllable parameters. This result quantitatively suggests temperature and time windows for maximum ethylene production which is the most valuable product, and for minimum the coke precursor aromatics formation.

This work provides quantitative guidelines to design stable NOCM system which avoids coke formation. The inclusion of accurate gas-phase kinetics allows for systematic investigation of how the ideal catalyst could perform in a NOCM system.

References

1. Zhang, H., Liu, G., Shi, L., and Ye, J. *Energy Mater.* 8, 1701343 (2018).
2. Guo, X., Fang, G., Li, G., Ma, H., Fan, H., Yu, L., Ma, C., Wu, X., Deng, D., Wei, M., Tan, D., Si, R., Zhang, S., Li, J., Sun, L., Tang, Z., Pan, X., and Bao, X. *Science*. 344, 616 (2014).
3. Chu, T., Buras, Z.J., Oßwald, P., Liu, M., Goldman, M., and Green, W.H. *Chem. Chem. Phys.* In press.

4. Goodwin, D.G., Moffat, H.K., and Speth, R.L. *Cantera: An object-oriented software toolkit for chemical kinetics, thermodynamics, and transport processes.* https://www.cantera.org, 2018. Version 2.4.0. doi:10.5281/zenodo.170284

567956 Catalyst Design for Microwave Assisted Methane Dehydroaromatization

The abundance of domestic natural gas reserves has resulted in increasing interests in generating higher value chemicals from methane, the main component of natural gas. While methane upgrading is currently achieved through an indirect route via syngas, direct upgrading offers conceptually a less energy intensive and costly alternative. One promising direct upgrading route is the non-oxidative conversion of methane to aromatics via dehydroaromatization (DHA). However, at the very high temperatures thermodynamically required for this reaction to proceed, carbon deposition becomes inevitable and results in rapid deactivation of the catalyst, rendering the process uneconomic. In the present project, we are exploring a novel process for methane upgrading by applying microwave radiation. Microwave radiation could provide a much more efficient way than conventional heating to couple energy into the reaction system, and is expected to do so more selectively by selectively heating the active (metal) site of the catalyst while leaving the bulk catalyst and gas phase at a lower temperature to avoid or reduce carbon formation. However, due to their complexity, microwave-assisted heterogeneous reactions are poorly understood to-date and systematic studies on how microwaves interact with catalytic materials are largely lacking. Here, we present results from an on-going study that aims to improve the understanding of catalyst design for microwave catalysis by systematically exploring and identifying catalyst properties that affect microwave absorption.

The catalyst for methane DHA is typically composed of metal nanoparticles dispersed in zeolites. In these catalyst systems, parameters including zeolite particle size, the concentration of Brønsted acid sites, speciation and dispersion of metals, and the nature of the metal can be expected to not only affect the catalytic reactivity but also to influence the ability of the catalyst to absorb and interact with microwave radiation. The main objective of our study is hence to obtain an improved understanding of what gives metal/zeolite-based catalyst higher microwave sensitivity. Towards this goal, we synthesized carefully defined metal/ZSM-5 catalysts to individually vary the size and Si/Al ratio of the zeolite ZSM-5, as well as speciation and type of metal in the catalyst. We find that incorporating more Al into the ZSM-5 framework increases the microwave sensitivity, while the size of ZSM-5 has a comparatively small effect. On the metal side, Fe-ZSM-5 was prepared via three different synthetic approaches to control metal dispersion and speciation. We find that both metal speciation and metal loading have significant impact on microwave sensitivity. Further studies regarding the nature of the metal are currently on-going. The observed effects will be presented and discussed in detail in the presentation.

Overall, through these studies, we aim to develop a fundamental understanding of metal/zeolite catalysts in microwave catalysis as a basis for rational catalyst design for microwave-assisted methane DHA.

573510 Microwave Assisted Catalytic Conversion of Ethane to Aromatics for a More Efficient Approach over a Conventional Fixed Bed Reactor

The mechanism of ethane dehydroaromatization over a 4%Mo-0.5%Fe/ZSM-5 catalyst was tested in a 6.650 GHz microwave fixed bed reactor (MW) and compared to a conventional fixed bed reactor (CFB). Over 80% ethane conversion was achieved under microwave irradiation at 375°C bulk temperature. The conventional fixed bed required a corresponding temperature of 660°C to achieve the same conversion as the microwave fixed bed reactor. The selectivity's of the catalyst were compared after 10 minutes of time-on-stream where the conversion of ethane was the same and reactant stream concentrations were fully developed. The CFB had a higher selectivity to aromatics; however, the MW presented a selectivity to ethylene of 26.5% higher than the CFB. The MW high selectivity to ethylene suggest that the aromatization reaction was limited possibly do to ethylene's inability to rapidly diffusion into the pores of the zeolite structure. This was attributed to the overall low bulk temperature of the catalyst bed or uneven heat distribution throughout the catalyst bed. TPO analysis of CFB showed three distinct orders of coke formation, whereas the microwave resulted in a single higher order temperature burring coke. Less total coke formation was observed on the MW spent catalyst, this was attributed to the low bulk temperature in the microwave which suppressed the formation of amorphous coke on the catalyst surface.

572117 Effects of Low Dopant Concentrations on the Performance of Samarium Oxide in the Oxidative Coupling of Methane

The oxidative coupling of methane (OCM) is a potential synthesis route to produce ethylene directly from methane, or natural gas, in one step. A number of different materials have been investigated in this reaction, but few catalysts display high activity, thermally stability and selectivity towards C_{2+} products (ethane, ethylene and higher hydrocarbons). Some of the most promising catalysts tend to be complex mixtures of oxides which makes identifying structure activity relationships challenging. To simplify the structure activity relationship in order to investigate fundamental effects, we have chosen to study the effects of low molar dopant concentrations on samarium oxide (Sm_2O_3) in the oxidative coupling on methane. We found that even at concentrations as low as 0.1 and 1.0 mol %, the dopants affect the methane conversion and the selectivity to ethane and ethylene of samaria with the highest increases in catalyst efficacy observed on 0.1% loading. Catalyst characterizations reveal that the dopants affect the crystal structure, as well as the number and strength of the basic sites, of samaria both of which are expected to influence the OCM properties.

569296 Enhancing Selectivity in Oxidative Coupling of Methane Using Catalytic Solid Oxide Membrane Reacting Systems

Oxidative coupling of methane (OCM) is a chemical process in which methane is converted directly into ethylene and ethane (C_2) at high temperatures (600-900°C) in the presence of oxygen and typically an active catalyst. The main challenge with this process is the formation of thermodynamically stable over-oxidation products, mainly CO and CO_2, which limits the C_2 selectivity. This selectivity problem is particularly significant in conventional reactors (co-fed with methane and oxygen) due to the presence of large gas-phase oxygen concentrations at the reactor inlet, which accelerates the unselective, deep oxidation reactions.[1] Catalytic solid oxide membrane reactors can minimize the unselective reactions by separating the methane and oxygen feed, and allowing the selective passage (conduction) of oxygen ions (O^{2-}) from the oxygen side to the methane side where the reaction occurs.[2,3] By allowing the distributed and controlled delivery of O^{2-} to the active sites, gas-phase oxygen can be significantly minimized, thereby limiting the occurrence of deep oxidation reactions. These membrane systems offer additional benefits including better control of reactor temperature and simultaneous separation of pure oxygen from other components in air thereby simplifying downstream separation. Herein, we have synthesized and tested catalytic solid oxide membrane systems in OCM. We show that a $BaCe_{0.8}Gd_{0.2}O_{3-\delta}$ (BCG) membrane system exhibits stable OCM performance and high carbon tolerance at extremely low oxidizing conditions. We also demonstrate that the BCG membrane system operated in membrane mode (separated and distributed oxygen feed) gives significantly higher C_2 selectivity compared to the same system co-fed with methane and oxygen, at comparable methane conversions. These results provide reliable experimental evidence of the improved OCM performance of a catalytic membrane system over a conventional co-fed system.

References

(1) Farrell, B. L.; Igenegbai, V. O.; Linic, S. *ACS Catal.* **2016**, *6* (7), 4340–4346.

(2) Farrell, B. L.; Linic, S. *Catal. Sci. Technol.* **2016**, *6*, 4370–4376.

(3) Igenegbai, V. O.; Meyer, R. J.; Linic, S. *Appl. Catal. B Environ.* **2018**, *230*, 29–35.

559199 Multi-Scale Simulation Framework for the Oxidative Coupling of Methane: From Process Layouts to Catalyst Particles

Introduction

The Oxidative Coupling of Methane (OCM) represents a direct valorization route for natural/shale/bio gas resources which has captivated both academic and industrial stakeholders for almost 40 years[1]. It is widely recognized that the main challenges of OCM, situated in difficult CH_4 activation combined with an inverse relationship between methane conversion and selectivity towards the desired C_2 (C_2H_6, C_2H_4) products and strong exothermicity, can be overcome only via simultaneous optimization of process layout, reactor design and catalyst composition[2]. In fact, the only claimed commercial implementation so far, the GEMINI process by Siluria and Linde[3], has introduced significant novelties in all these three fields.

Nevertheless, most of the OCM process studies currently available in literature are based on a single catalyst and a single reactor configuration. The performance indicators at the outlet of the OCM reactor are 'imposed'[4] or simulated with simplified kinetics[5]. Variations in CH_4 conversion and C_2 selectivity with the process conditions or catalyst composition would have a dramatic impact on the separation and recycle sections downstream and cannot be easily extrapolated to other catalysts.

In order to overcome these limitations, in the present work a multi-scale simulation framework is proposed which enables process modelling for OCM by incorporating a detailed reactor model and a fundamental kinetic model specifically accounting for the catalyst properties.

Simulation framework

A schematic representation of the simulation framework developed is shown in Figure 1. Three different modelling challenges were tackled and are displayed in the figure:

Process model

Process simulations were performed via the commercial software Aspen Plus. An example of a potential process flowsheet[6] is shown in Figure 1/A. The major challenge of the present work lies in the detailed modelling of the reaction section. This was tackled by coupling Aspen Plus to an external, in-house code. The implementation occurred via a customized user unit operation model[7] (Figure 1/B1), namely a Fortran interface which is capable of: reading input from Aspen streams and block variables, write input to the external code, execute the external code, read output from the external code, write output to Aspen streams.

Reactor model

The external code solves mass, energy and momentum balances for a catalytic fixed-bed reactor (Figure 1/B1), which can be operated both isothermally and adiabatically[8]. The reactor model is 1-dimensional, i.e. no radial gradients are considered on the reactor scale, and heterogeneous, i.e. irreducible concentration and temperature gradients on the particle scale are taken into account. The two phases considered on the particle scale, and indicated in Figure 1/B2, are: the *intraparticle* phase, consisting of the catalyst particles and the gas contained in their pores, and the *interstitial* phase, accounting for the gas around the particles. Textural properties of the catalyst particles (surface area, porosity, tortuosity, diameter) are included among the model parameters.

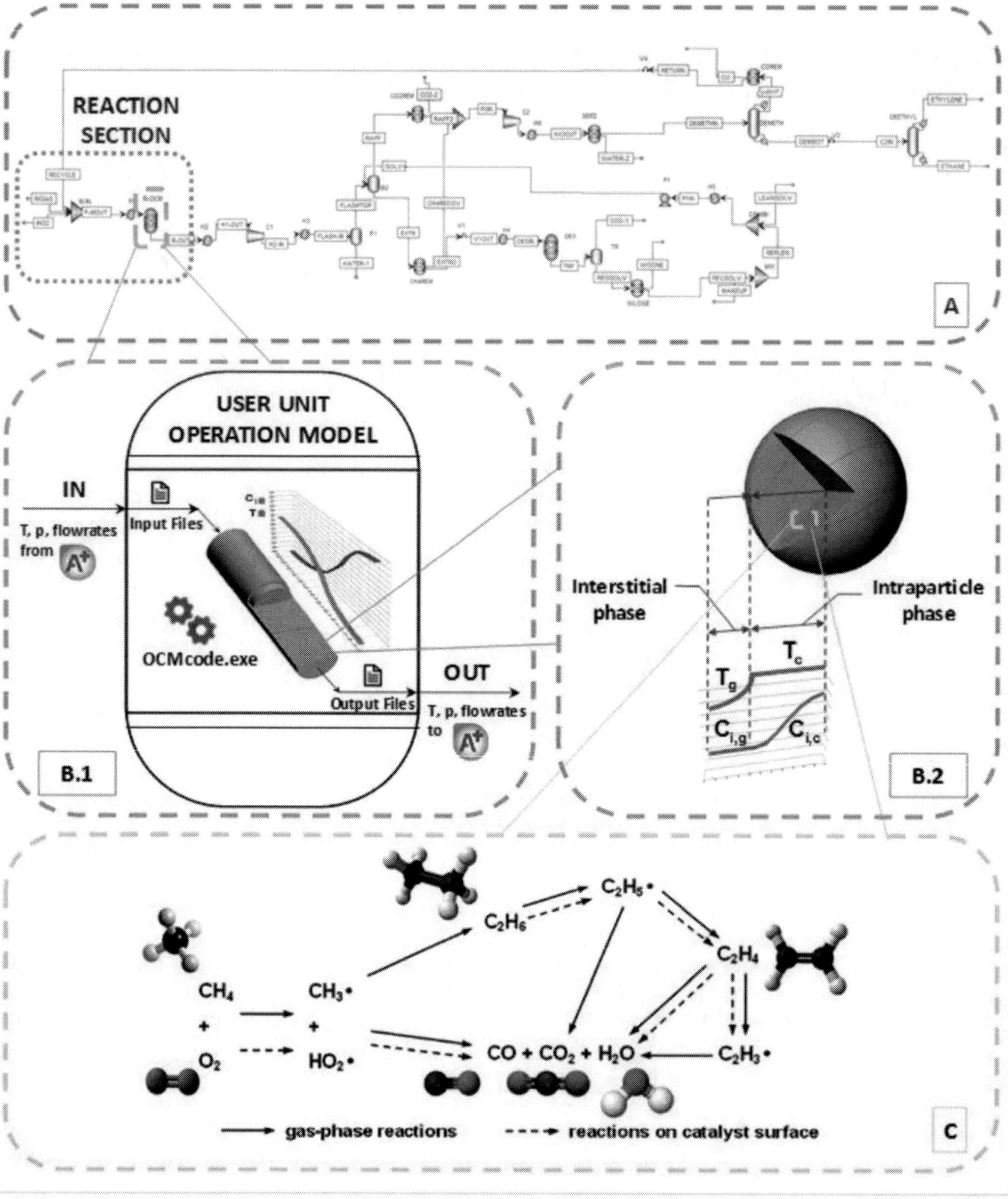

Figure 1. Multi-scale simulation framework deployed in the present work. A. Example of process flowsheet for OCM[6]; B. Customized user unit operation model, enabling the simulation of an OCM reactor (B.1), including particle-scale phenomena (B.2)[8]; C. Simplified representation of the reaction scheme[9].

Microkinetic model

A microkinetic model for the OCM reaction, previously developed in our research group and validated for five different catalysts[9], was embedded in the reactor model. The model accounts for 13 molecules, 11 radicals and 11 surface species, and comprises 39 reversible gas-phase reactions and 26 reversible reactions on the catalyst surface. A simplified representation of the reaction scheme is reported in Figure 1/C. The possibility of reproducing the performances of different catalysts arises from the presence in the model of catalyst descriptors, i.e. kinetically significant catalyst features - such as chemisorption enthalpies, sticking coefficients, density of active sites - that specifically account for the impact of the catalytic material on the reaction kinetics.

Case study: distributed oxygen feed and intermediate cooling

In this case study, the reaction section of a potential OCM plant was simulated considering a Sn-Li/MgO catalyst[9]. The base case, corresponding to scenario A.I in Figure 2, was characterized by a single fixed-bed reactor operated adiabatically, at atmospheric pressure, with a methane-to-oxygen molar ratio equal to 10. The space time was set to 17 kg s/mol$_{CH4,0}$ via a design specification to achieve complete oxygen conversion at the reactor outlet. The inlet temperature was chosen equal to 840K via a design specification to operate the reactor with maximum temperature equal to 1273K[8]. The results obtained in terms of CH_4 conversion, C_2 selectivity and C_2 yield are reported in the first column of Table 1.

A second scenario, indicated with A.II, was simulated by considering the same process conditions as A.I, but reducing CH_4/O_2 to 5. As highlighted in the second column of Table 1, the resulting outlet temperature was way beyond the limit of 1273K, due to the higher adiabatic temperature rise associated to the higher oxygen content. This scenario exemplifies that not all sets of process conditions are possible for the operation of an adiabatic OCM reactor[8], and, hence, was discarded.

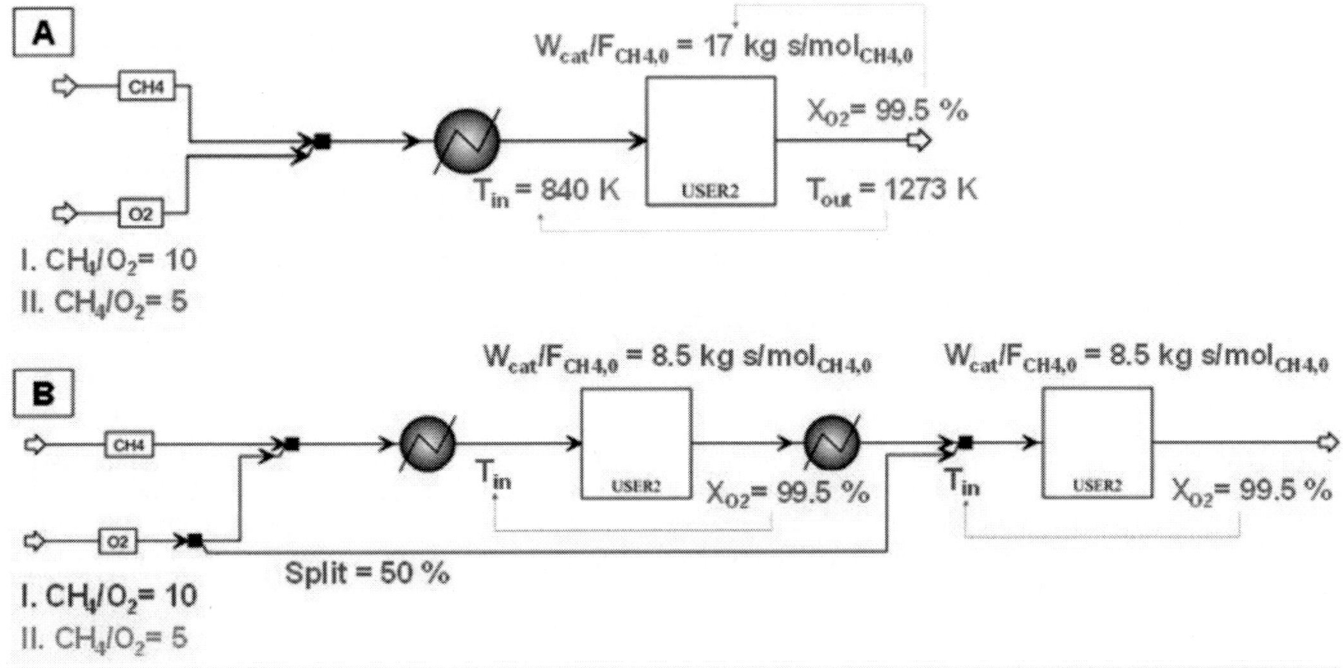

Figure 2. Process layout of the OCM reaction section in the present case study, based on Sn-Li/MgO catalyst. The values of CH_4/O_2 reported are referred to the global methane and oxygen molar flowrates.

Scenario B.I was characterized by two Sn-Li/MgO catalyst beds in series, with distributed oxygen and intermediate cooling. The amounts of CH_4, O_2 and catalyst were kept constant in comparison with scenario A.I, but catalyst and oxygen were split in two, with 50% of the oxygen being fed to the first bed, containing half of the catalyst, while the other 50% was fed to the second bed. The inlet temperature to each bed was adapted via a design specification, again to achieve complete oxygen conversion at the reactor outlet. As it can be observed in the corresponding column in Table 1, this scenario resulted in higher inlet temperatures. This is due the fact that catalytic light-off is more difficult to achieve when methane is present in high excess[8]. The corresponding outlet temperatures were lower for both beds, due to the limited methane conversion that was obtained with such a low oxygen content. This resulted in a lower C_2 yield as compared to the base case.

Scenario B.II was analogous to B.I, this time considering CH_4/O_2= 5. In Table 1 can be observed that, compared to the base case, this process configuration resulted in an increase of +22% in the yield. The price to pay was +40K in the required inlet temperature and -19% decrease in the selectivity.

As the computational time for each simulation was in the order of 30s, the present case study could easily serve as basis for further techno-economic analyses aimed at evaluating the profitability of different configurations, such as scenarios A.I (higher selectivity) and B.II (higher yield), embedded in a full process scheme, including also the separation and the purification sections and the recycle of material streams.

Table 1. Simulation results for the present case study. For all the scenarios: p= 1bar, $W_{cat}/F_{CH4,0}$= 17 kg_{cat} s/mol_{CH4}. The variables of major interest are indicated in bold.

	Scenario A		Scenario B (multiple beds with intermediate cooling)					
	I	II	I			II		
N. of beds	1	1	2			2		
			Global	First Bed	Second Bed	Global	First Bed	Second Bed
CH_4/O_2	10	5	10	20	18	5	10	9
T_{in}	840K (manipulated)	840K	-	883K (manipulated)	861K (manipulated)	-	880K (manipulated)	855K (manipulated)
T_{out}	1273K (controlled)	1425K		1125K	1111K	-	1273K	1263K
X_{O2}	100% (controlled)	100%	100%	100% (controlled)	100% (controlled)	100%	100% (controlled)	100% (controlled)
X_{CH4}	15%	31%	12%	8%	4%	22%	14%	20%
S_{C2}	75%	64%	66%	77%	46%	63%	76%	40%
Y_{C2}	11%	20%	8%	6%	2%	14%	11%	4%

The few variations investigated in this case study were mainly aimed at demonstrating the potential of the developed simulation framework and more scenarios are currently being pursued. For instance, ongoing simulations evaluate how activities and selectivities of alternative catalysts, such as Sr/La_2O_3 and $NaMnW/SiO_2$, impact the optimal process configuration identified from yield and/or selectivity maximization.

Conclusions

The simulation framework developed in the present work enables multi-scale modelling of the OCM process, without compromising on transport phenomena, both on reactor and particle scale, and kinetics, which is herein described in detail via a microkinetic model. The enablement of process simulations based on microkinetic models represents a step forward in the design and assessment of realistic process layouts for natural gas valorization. In fact, the possibility of evaluating the impact of process conditions, feed composition and catalyst properties on the outlet stream from the reaction section paves the way for more reliable techno-economic assessments aimed at designing an industrially-viable OCM process.

References

1. Keller, G.E. and M.M. Bhasin, Journal of Catalysis, 1982

2. Kondratenko, E.V., et al., Catalysis Science & Technology, 2017

3. https://www.linde-engineering.com/en/process_plants/petrochemical-plants/oxidative-coupling-methane/index.html

4. Ortiz-Espinoza, A.P., et al., Computers & Chemical Engineering, 2017

5. Godini, H.R., et al., Fuel Processing Technology, 2013

6. Arickx, M., et al., Internal Report, 2018

7. Tripodi, A., et al., Catalysts, 2017

8. Pirro, L., et al., Industrial & Engineering Chemistry Research, 2018

9. Alexiadis, V.I., et al., Applied Catalysis B, 2016

560651 Single Site Vs Crystalline: Development and Characterization of Mn/W/Na/SiO2 Model Catalysts for the Oxidative Coupling of Methane (OCM)

An abundance of natural gas due to the advancements in exploration and extraction technologies has rejuvenated interest in the direct conversion of natural gas to ethylene via oxidative coupling of methane (OCM). The tri-metal oxide supported Mn/W/Na/SiO2 catalyst has been studied in the past and has shown ~75% C_2 selectivity, ~35% conversion and robust thermal stability under operating conditions. However, improvements need to be made to the catalyst to increase product yield and lower the operating temperature. Most published studies of the tri-metal oxide catalysts before and after OCM typically identify crystalline phases such as Na2WO4, MnWO4, etc. as the active phases. Our in-situ and operando Raman spectroscopy studies of supported Mn/W/Na/SiO2 catalysts during OCM, surprisingly, reveal that the often suggested crystalline phases are not present during reaction and that dispersed surface metal oxide species are present. This suggests that the surface metal oxide species may be the catalytic active sites for OCM by supported Mn/W/Na/SiO_2 catalysts. Developments regarding design and performance of dispersed and crystalline tri-metal oxide catalysts for OCM at/near working conditions will be presented in this work. The focus will be on *in-situ*, and *operando* Raman spectroscopic characterization of the model tri-metal oxide Mn/W/Na/SiO_2 catalysts to establish the nature of the surface active sites, their H_2-TPR characteristics, and their structure dependent catalytic performance.

579675 U.S. Department of Energy Fossil Energy R&D Program: Overview and Priorities

The Office of Fossil Energy (FE) within the U.S. Department of Energy (DOE) has identified five overarching research and development (R&D) priorities that require concentrated federal investments to maintain America's dominance in the fossil energy sector: 1) advance small-scale modular coal plants of the future, 2) improve the performance, reliability, and efficiency of the existing coal-fired fleet, 3) reduce the cost of carbon capture, utilization and storage to enable wider deployment, 4) optimize recovery of oil and gas resources with real-time analysis and machine learning, and 5) improve the efficient use of the Nation's scarce water resources.

This presentation will provide an overview of how DOE's FE R&D Program is addressing these five key R&D priorities, as well as related FE research needs articulated by the U.S. Congress. Emphasis will be placed on potential future R&D thrusts that will require chemical engineering solutions.

The presentation will also explain the approach taken by the government-owned, government-operated National Energy Technology Laboratory to mature and commercialize technology for DOE/FE, and explain how academia and industry can partner with DOE to conduct R&D through grants and cost-shared financial assistance.

579673 Energy Conversion and Utilization in Carbon-Constrained World: Carbon Capture, Utilization and Storage (CCUS)

While more than 2 billion people have been working themselves out of poverty, world energy demand has been growing rapidly. Fossil fuel resources are not running out, but their use produces carbon dioxide (CO_2), which, as the most important greenhouse gas, drives climate change. At COP21, held in 2015, Mission Innovation (MI) – a global initiative of 23 countries and the European Commission (on behalf of the European Union) – was launched to reinvigorate and accelerate global clean energy innovation with the objective to make clean energy widely affordable. Eight innovation challenges, including Carbon Capture, Utilization and Storage (CCUS) have been selected. There have been significant efforts to develop CO_2 capture materials and some (e.g., amine-based aqueous solvents) have been demonstrated in large scales. Further, a number of CO_2 injection projects have shown promising results. Unfortunately, the energy penalties associated with various steps of CCUS processes are still considered to be high. This talk will discuss the perspectives, challenges and opportunities of novel CCUS technologies and the recent scientific advancements have been made in this field.

579674 Building Digitally Coupled Learning and Innovation Processes

Companies struggle to find people with the necessary skills for today's economy, from truck driving to data mining. Apart from the talent crisis, companies must reduce new product cycle times and then deliver improved innovations to market, faster. Workers in every industry, meanwhile, find it more and more difficult to climb a career ladder, or even to get onto the bottom rung. For learners of all ages, the rapid evolution in technology and business practices accelerates and intensifies the need for new approaches to career awareness, preparation, and retraining. To help address both human capital development and the need for new knowledge universities and businesses must innovate and think "outside the box" to create new engagement models that improve quality, create customer value, provide a hands-on educational experience and build a foundation for continuous learning. The AIChE has an opportunity to build on existing practices to design and implement new learning and innovation practices. This presentation will cover background and energy application examples under development.

568677 Effect of Synthesis Methods on Dry Reforming of Methane over Ni/$Ce_xZr_{1-x}O_2$

Greenhouse gases are responsible for global warming and recent climate changes. Dry reforming of methane has drawn much attention as it converts two greenhouse gases (CH_4 and CO_2) to syngas with a H_2/CO ratio of 1. The reaction is attractive for eliminating greenhouse gases, exploiting natural gas resources with high CO_2 content and producing syngas for at H_2/CO ratio closer to unity for further chemical production. Ni-based catalysts have shown great activity towards dry reforming of methane. Unfortunately, such catalysts often suffer from severe coking and sintering during the reaction which leads to fast deactivation of the catalysts and prevents the industrial application of the reaction.

To address the challenge, our group designed and synthesized a series of Ni/$Ce_{0.82}Zr_{0.18}O_2$ (Ni/CZ) that were found to activate methane at low/moderate temperatures. Synthesis methods applied include dry impregnation (DI), strong electrostatic adsorption (SEA), co-precipitation (CP) and combustion synthesis (CS). Catalysts with different nickel loading were synthesized to tune the accessibility of surface nickel particles for bulk synthesis methods (CP and CS). XRD analysis showed that only the CP catalysts with 6 and 10 wt.% contained detectable nickel particles. EXAFS analysis on fresh samples confirms the presence of small and well-dispersed nickel particles on all catalysts. For 2CS and 2CP, almost atomic dispersion of nickel was achieved.

Reactivity tests showed that all catalysts were active for dry reforming of methane at 700 °C to the thermodynamic limit. The catalyst remains active after 12 hours of experiment. Carbon deposit production occurred only within the first hour of the reaction. TGA of spent catalyst showed that a limited amount of carbon deposits formed within 12 hours of reaction. XANES spectra of the spent catalysts showed that except for the DI catalyst, all nickel stayed reduced after the reaction, indicating carbonaceous deposits might prevent re-oxidation during sample handling in air. To further differentiate the performance of each catalysts, reactivity tests were performed with a diluted catalyst loading. The results showed that catalysts with smaller and better dispersed nickel clusters have better activity and are more resistant to deactivation.

573356 Reforming of Methane over Ni and Ca Based Pyrochlore Catalysts to Produce Syngas

Catalytic conversion of methane and carbon dioxide to syngas is the most studied and feasible option to utilize these gases. Further, syngas can be converted to high value chemicals. Some of the processes involving conversion of CH_4 and CO_2 to syngas are Dry Reforming, Oxy-CO_2 Reforming and Bi-Reforming of methane. In the current work, we focus on the former two, Dry reforming (DRM) and Oxy-CO_2 reforming of methane (OCRM). Nobel metals are most active and stable for reforming reactions, but, are expensive. Hence, comparatively inexpensive Ni based catalysts, which are significantly active and stable, are explored for these reactions. However, these catalysts are prone to deactivation by coking and sintering at high temperatures. Pyrochlores are thermally and chemically stable crystalline metal oxides with general formula $A_2B_2O_7$ which are tolerant at reforming conditions. In our work, Lanthanum Zirconate pyrochlore ($La_2Zr_2O_7$) was doped with Ni in Zr site and Ca at La site to produce an active and stable catalyst. A modified Pechini method was used to produce 3wt% Ni loading of $La_2Zr_2O_7$ catalyst (LCNZ3) and tested for reforming reactions. 1:1 ratio of CH4:CO2 was used for DRM, while 1:3:1 ratio of CO2:CH4:O2 was used for OCRM. Additionally, spectroscopic and thermal characterizations were carried out to study the catalyst.

XRD of fresh catalyst showed peaks corresponding to the base pyrochlore (La_2Zr_2O
NiO. Additionally, a perovskite phase was also observed in the catalyst. A significant increase in the lattice constant value of the pyrochlore was seen, suggesting Ni substitution into the pyrochlore structure. Ni^{+2} and Ni^{+3} states were observed to be present on the surface from XPS. XPS and SEM-EDS analysis were done to compare the surface and bulk compositions of the catalysts. TPR showed reduction peaks at different temperatures, indicating different interaction strengths of Ni in the structure. Catalyst showed higher activity, i.e., higher conversion values of reactants for OCRM compared to DRM. Also, DRM deactivated the catalyst more significantly compared to OCRM. This could be due to limitation of carbon formation on the surface because of presence of additional oxidizing agent, i.e., oxygen. Post run TPO were also carried out to characterized the coke deposited on the surface. Carbon deposited was observed to be present in more than one phase, i.e., amorphous and polymeric.

579280 Hydrogen Generation Using Heavy Oil Feedstocks

TDA Research has developed a patented process to steam reform heavy hydrocarbon fractions from petroleum such as atmospheric and vacuum residuum over nickel-based steam reforming catalysts without catalyst deactivation (allowed application 14/698,541, "Process for Generating Hydrogen from Heavy Oil or Hydrocarbons"). The catalyst is not deactivated because it is continuously regenerated with air, which burns off coke. In an industrial process with circulating fluidized beds (CFB), the catalyst is continuously regenerated, but in our laboratory experiments, the fluidized bed of catalyst was regenerated after about 90-120 minutes of steam reforming by first purging the Incoloy 800H reactor with nitrogen (to eliminate residual fuel) and then switching to a preheated air feed. The laboratory scale steam reforming and catalyst regeneration are done at 50 psig and 850°C (1562°F). In this sense our process functions somewhat like a chemical looping steam reformer (CLR).

571857 Deactivation of Unsupported and Silica-Supported Cobalt Nanoparticles for Ethanol Steam Reforming and CO_2 Methanation

Deactivation of unsupported and silica-supported cobalt nanoparticles for ethanol steam reforming and CO_2 methanation

Paola Riani,[a,b] Gabriella Garbarino,[b,c] Tullio Cavattoni[d], Guido Busca[b,c]

[a] *Università degli Studi Genova, DIFAR - Department of Pharmacy, Genoa, Italy*

[b] *INSTM, UdR Genova, Genoa, Italy*

[c] *Università degli Studi Genova, DICCA - Department of Civil, Chemical and Environmental Engineering, Genoa, Italy*

[d] *Università degli Studi Genova, DCCI - Department of Chemistry and Industrial Chemistry, Genoa, Italy*

**Corresponding author: Guido.Busca@unige.it*

1. **Introduction**

Supported Cobalt catalysts, in particular Co/Al_2O_3 and/or Co/ZrO_2-SiO_2, are used industrially for the low temperature Fischer Trøpsch process producing Diesel fuel and waxes from syngases [1]. In this reaction, methane is actually the main product on molar basis. Some studies report on a remarkable activity of cobalt catalysts also for methanation of CO and CO_2 (CME) [2] while others [3] ranked low cobalt catalyst for these reactions for low activity, selectivity and stability. Cobalt catalysts are also of interest for Ethanol Steam Reforming (ESR), a potentially useful process for producing renewable hydrogen from bioethanol.

Concerning catalytic activities, studies on these reactions sometimes suggest that the support plays an important role in the activation of reactants. However, we previously found that both cobalt [4] and nickel [5] and their alloy [6] may be very active for ESR also as unsupported metal nanoparticles (NPs), and as NPs supported on inert carriers [7], suggesting that the support is not necessary for reactant activation for this reaction. However, the same Ni NPs were found not active in the CME reaction.

To go deeper in our studies, we prepared unsupported and silica – supported cobalt NPs starting from different precursors and with different procedures, and we tested them in both CME and ESR. The aim was to reveal the role of preparation method and support on activity and short-term stability in these two reactions.

2. **Experimental**

Data on the preparations of the catalysts under study are reported in the Table.

Notation	Support	Metal load	Precursor	Reductant	Preparation/ Separation
A	--	¥	Chloride	$NaBH_4$	filtration
B	--	¥	Nitrate	$NaBH_4$	Filtration/sonication

C	--	¥	Carbonyl	--	Filtration
D	SiO_2 gel	20 %	Acetate	--	Wet Impregnation
E	SiO_2 gel	20 %	Nitrate	--	Wet impregnation

Catalysts have been characterized by XRD, UV-vis-NIR, IR (skeletal), FE-SEM, before and after reaction. Catalytic reactions were realized in continuous fixed bed reactors. Product analyses were performed using GC, GC-MS and on-line IR.

3. Results and discussion.

Catalysts A and B revealed essentially no activity in CME in the Trange 523-773 K, although they produced some amount of CO by reverse Water Gas Shift (r-WGS). However, catalysts became rapidly deactivated also with respect r-WGS activity. Characterization data suggest that low activity is associated to the presence of impurities arising from the precursor, in particular due to boron contamination.

In contrast, catalyst C showed significant activity in CME in the temperature range, with selectivity to methane higher than to CO, but with an evident deactivation of CME upon time on stream at any temperature between 523 and 773 K. On the other hand, the catalyst showed nearly constant r-WGS activity upon time on stream, producing CO. After using it at 773 K the catalyst was almost completely deactivated with respect to CME while showing constant activity towards r-WGS, maximum yield to CO being 45 % at 773K. A very similar behavior was found for supported catalyst D, which also showed initial CME activity higher than r-WGS activity, the former however affected by fast deactivation, while the latter is nearly stable. In both cases FE-SEM data indicate that deactivation is mostly associated to the formation of encapsulating carbon. Catalyst E instead shows significant activity in CME producing up to 67% methane yield at 673 K, with the coproduction of lower amounts of CO. The activity is affected by very slow deactivation phenomena in this case, apparently associated to sintering of NPs. Both cubic and hexagonal Co is found in all spent catalysts. The lower activity and lower stability towards methanation of sample D with respect to sample E (both 20% Co/SiO_2) can be tentatively attributed to the larger cobalt particle size in the former. Thus, smaller particles are more active in methanation and less subjected to deactivation by encapsulating carbon. In any case, the r-WGS activity producing CO seems to follow a parallel and independent path with respect to CME. This activity is attributed to CoO and/or Co-silicate species evident in the spent catalysts.

All catalysts under study are active in ESR, showing that boron impurities do not hinder such a reaction. All catalysts are highly selective for ethanol dehydrogenation to acetaldehyde at 523-673 K, while produce ESR predominantly at T > at 673-773 K. This suggests that high temperatures are needed for the activation of water. NPs produced from reduction with $NaBH_4$, thus containing boron impurities, may be even more selective to hydrogen (yields > 85 %) perhaps because boron impurities may selectively poison the sites for producing methane by acetaldehyde cracking.

The active phase for ESR is represented by cubic cobalt, whose ESR activity is greater for smaller particles. Slow deactivation is mainly associated to sintering. However, these particles are also active in producing carbon nanotubes. On the other hand, the formation of carbon nanotubes does not hinder catalytic activity at the timescale of a few hour laboratory experiment.

Depending on the precursor, the catalyst may deactivate by forming larger Co-containing nanoparticles embedded on the silica/cobalt silicate surface. These particles, possibly containing also silicate species, are also not active in producing carbon nanotubes, i.e. they are unable to both activate water and favor C-C bond formation.

The data reported above show that support is not needed for catalytic activity in CME and ESR but may be important for catalyst stability. However, the silica support may also favor catalyst deactivation of cobalt NPs.

References.

[1] Busca, G.; Heterogeneous Catalytic Materials, Elsevier, **2014**, pp. 314-315.

[2] Le, T.A.; Kim, M.S.; Lee, S.H.; Park, E.D.; Top. Catal. **2017** 60, 714–720

[3] Kuznecova, I.; Gusca, J.; Energy Procedia **2017,** 128**,** 255–260

[4] Riani, P.; Garbarino G.; et al. Appl. Catal. A: Gen. **2016**, 518, 67–77

[5] Riani, P.; Garbarino G.; et al. J. Mol.Catal. A: Chem. **2014,** 383–384, 10–16

[6] Garbarino, G.; Cavattoni, T.; et al. Catal. Lett. **2019,** 149, 929-941

[7] Riani, P.; Garbarino G.; et al. J. Chem. Techn. Biotech. **2019**, 94, 538-546

572017 Toluene Reforming Using Nickel Based Catalysts Made from Mining Residues

The formation of aromatic compounds is one of the problems facing the processes of gasification and pyrolysis. The usage of nickel-based catalysts is a prospective method for converting tar. These catalysts deactivate quickly due to sintering when carbon is formed. The patent-pending catalyst Ni-UGSO made from mining residues is used in this study for tar reforming. Ni-UGSO showed high activity and high resistance to carbon deposition in methane and bio-oil reforming.

Toluene, which forms approximately 24% of tar is used as tar model compound. The catalyst is prepared by the incipient wetness method through mixing the mining residues with nickel nitrates, then drying and calcining the mixture for 12 hours at 900°C. The reforming runs were conducted in a fixed bed reactor. Toluene and water were emulsified with H_2O/C ratio equal to one.

The tested parameters are:

- Temperature from 650 to 850°C, to determine the activation temperature of the catalyst.
- Variation of Nickel mass fraction, to test its influence on the total conversion, yields of H_2 and CO and carbon formation.

GC analysis of the product gases showed that the main products were H_2 and CO, with 60% and 75% yields, respectively, while the overall conversion was between 85 and 98%. At the locally optimal conditions, a 24h long test was performed. During this period the fraction of carbon deposited on its surface was about 2% of the total carbon injected in the reactor, without signs of deactivation.

The activity of the catalyst was then tested by adding naphthalene to the emulsion; the reforming results revealed high decomposition of the mixture with a total conversion of 73%.

The same reforming experiment was repeated using another catalyst made from olivine and activated with nickel to compare its efficiency with that using Ni-UGSO.

All tested catalysts were analysed before and after the reforming experiments through XRD, BET and FEG/SEM microscopy. These analyses provided useful information on the involved crystalline phases, the specific surface area changes as well as the changes of the catalyst's particles morphology.

567347 Low Temperature Steam Methane Reforming Using Lanthanides Modified Ni/Al_2O_3 Catalyst for Hydrogen Production for Fuel Cells

A series of selected lanthanides (La, Ce, Pr, Sm, Eu, and Gd) modified Ni/Al_2O_3 catalysts were prepared by impregnation technique for low temperature methane steam reforming to produce hydrogen for fuel cells. According to the thermodynamic equilibrium analysis, it is possible to operate methane steam reforming at low temperatures. Low temperature steam reforming in combination with hydrogen selective membrane presents great potential of intensifying the classical industrial hydrogen production process via natural gas. A highly active catalyst is required for the production of hydrogen at lower temperatures. The catalysts prepared in this study were characterized by X-ray diffraction, Brunauer–Emmett–Teller (BET), transmission electron microscopy (TEM), scanning electron microscopy, energy dispersive X-ray, and temperature-programmed reduction (TPR) characterization techniques. The performance evaluation of the prepared catalysts was conducted in a fixed bed micro-reactor at steam to carbon ratio of 3.0 and at a pressure of 2.1 bar over a range of temperature 500 to 600 C. The results indicated that Pr modified Ni/Al_2O_3 catalyst performed better in terms of conversion of methane to other catalysts in the series. Further investigations were carried out by varying the amount of Pr over a range of 1 to 7 wt% and found that 3% Pr modified 12% Ni/Al_2O_3 catalyst showed better performance for the SMR reaction. Pr–Ni catalyst has the higher BET surface area among other synthesized lanthanide-promoted 12%Ni/γ-Al_2O_3 catalysts. TEM analysis indicated uniform metal dispersion of promoted catalysts. Moreover, TPR analysis indicated an optimum interaction between the metal and the support in the Pr–Ni catalyst. The results on catalyst screening, testing and characterization will be presented along with follow-up reactor design for integrated membrane reactor with reaction and separation for hydrogen production.

557168 Autothermal (CO_2 and O_2) Di-Reforming of Methane with the Ni-UGSO Catalyst: Lab-Scale Results and Kg-Lab Scale Tests at Industrial H_2 Production Conditions

This work belongs to a larger endeavor aimed at studying the performance and the stability of a patent pending hydrocarbons reforming spinellized nickel catalyst prepared from an ilmenite metallurgical residue. Said residue is a negative value upgraded slag oxide (UGSO) coming from a TiO_2 slag production unit, operated by Rio Tinto Iron & Titanium, Quebec, Canada. Recently published results from experiments carried out with this catalyst exhibited high hydrogen production selectivity during methane steam, dry (CO_2) and autothermal (CO_2 and O_2) reforming and stability over time-on-stream (TOS).

At the lab-sale barometric conditions the best results are registered at T=850°C, molar ratios of $CH_4/O_2=2$ and $CH_4/CO_2=3$ and space velocity of GHSV=4500+/-100$ml_{STP}/(h.g_{cat})$. In these conditions, the catalyst remained stable without any deactivation and nil carbon formation during the entire 2 days of operation. The CH_4 conversion (98%) and the H_2 and CO yields (respectively 98.8% and 95.5% were equally stable and near-thermodynamic equilibrium. The steady state operation is characterized by the coexistence of multiple phases in the catalyst structure, including metallic, oxides and spinels.

Based on these results, a new project is funded by the Quebec and Canada Governments as well as our industrial partner (Rio Tinto Iron and Titane). Said project focuses on scaling up the process by a factor of 1000 (from g-lab to kg-lab scale); it aims at proving that both the catalyst (Ni-UGSO) and the autothermal direforming process can compete with the prevailing catalysts and processes for the production of H_2 from natural gas steam reforming. The main challenges are to prove that:

- this much cheaper catalytic formulation has at least the same efficiency, robustness and life cycle characteristics with the best commercially available catalysts;
- the autothermal direforming proposed is at least equally effective as the presently only commercially available process for H_2 production, that is ‘steam reforming’.

In this publication, we will present (a) the lab-scale results and the mechanistic explanation of the catalyst efficiency and robustness; (b) the scale-up calculations and criteria used; (c) the kg-lab set-up and operation protocol and (d) the first results at kg-lab scale towards answering the above challenges.

569286 Recent Progress on Key Technical Barriers for Chemical Looping Combustion

Chemical Looping Combustion is an environmentally sustainable approach for converting abundant resources of fossil fuels. For the last five years, the U.S. Department of Energy's National Energy Technology (NETL) has been investigating this concept as an approach to mitigate greenhouse gas emissions in future power plant concepts. This presentation will review the progress in the chemical looping combustion technology development over the last five years with a primary focus on internal research and development efforts completed by NETL's Research and Innovation Center. The topic areas that will be discussed include the following:

- Results from plant simulations that identify key technology barriers for this technology
- Progress in oxygen carrier performance and durability which includes natural hematite, promoted hematite, and manufactured oxygen carriers
- Recommendations for advancing the technology readiness level for Chemical Looping Combustion in the future.

575572 Particle Properties in Techno-Economic Analysis of Chemical Looping Hydrogen Generation

Hydrogen (H_2) is considered as one of the most extensively used chemical intermediates in industry and a potentially attractive option for clean energy generation. However, CO_2 emissions from conventional H_2 production plants are significant, challenging the life cycle advantages of H_2 utilization. Thus, technologies are necessary to reduce or eliminate the carbon emissions associated with H_2 production while meeting the market demand in the present and future applications. Chemical looping hydrogen generation (CLHG) technology uses natural gas to generate H_2 with inherent CO_2 capture, which avoids contact between fuel and air (or oxygen) by utilizing a solid oxygen carrier to enable effective separation of O_2/N_2, CO_2 and H_2. CLHG can efficiently generate high purity H_2 with a high cold gas efficiency without the need for an amine-based CO_2 capture unit as required in conventional steam-methane reforming (SMR) or partial oxidation (POX) processes. In the CLHG process, iron-based metal oxide particles are circulated through a counter-current moving bed reducer reactor for full natural gas combustion to CO_2 and H_2O, a counter-current moving bed oxidizer reactor for reduction of steam by the reduced iron oxide to generate H_2, and a fluidized bed combustor reactor with riser for particle regeneration and transportation by air. After water removal, CO_2 can be separated from reducer flue gas and we can obtain high purity H_2 from oxidizer gas outlet. While particle is the key component of CLHG technology, this work focuses on the influence of particle properties, such as size, density, heat capacity, etc., on the overall CLHG process economics. Parametric studies of operating conditions were conducted to achieve auto-thermal operation of chemical looping reactor system and the pneumatic transportation of particles from combustor to reducer while varying particle properties. Specifically, this study developed the overall model of CLHG process with Aspen Plus simulation software. Several cases with representative particle properties and high process efficiency were configured. Energy and material balance information generated by the model was utilized to perform techno-economic analysis on the configured CLHG cases. The results indicate the requirement of particle properties to gain benefits of the CLHG system as compared to conventional H_2 generation process with CO_2 capture.

569092 High-Purity Hydrogen from Biogas with a Combined Reformer and Fixed-Bed Chemical Looping System

High-Purity Hydrogen from Biogas with a Combined Reformer and Fixed-Bed Chemical Looping System

The decentralized allocation of high-purity hydrogen plays a crucial role for a future hydrogen-based economy. Hydrogen will act as secondary energy carrier and temporary storage between fluctuating renewables as primary energy and the volatile demand from private and commercial consumers. The RESC process is proposed as a combination of a steam reformer and a fixed-bed chemical looping system within a single, compact reactor system for hydrogen generation by Hacker [1]. Depending on the syngas generation (i.e. reforming, gasification), different renewable and fossil feedstocks can be utilized. High pressure hydrogen generation was demonstrated in smaller lab systems reaching a hydrogen release pressure of up to 100 bar by Zacharias et al. [2].

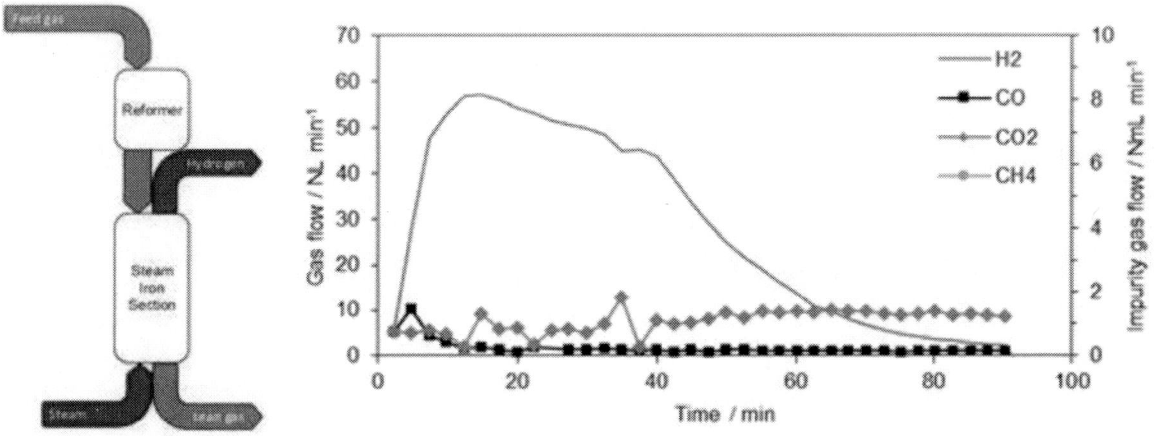

Figure 1: Schematic of RESC process (right) and exemplary hydrogen product gas in oxidation phase (left).

As part of the research activities, an experimental plant capable of hydrogen generation with an output power of 10 kW (LHV-based) from methane and 18 kg total inventory of oxygen carrier was presented by Bock et al. [3]. The system was capable of high purity hydrogen generation exceeding 99.999% from methane feedstock with carbon monoxide and carbon dioxide being the main contaminants. Ongoing research especially focuses on the utilization of biogenic resources as biogas or gasified biomass for hydrogen generation. Within the experimental research, the applicability of synthetic biogas mixtures was proved by Bock et al. [4]. Therefore, several biogas compositions were investigated regarding their suitability for hydrogen generation. The results confirm the high potential of biogas as potential source for high-purity hydrogen generation between 99.99%-99.999% purity as demanded by low temperature fuel cells (see Figure 1). Nevertheless, the application of synthetic biogas also led to increased carbon contamination compared to earlier experiments with methane as feedstock. The degree of utilization for hydrogen generation from synthetic biogas exceeded 40% within the experimental series, depending on the biogas composition. In addition, a variation of the process temperature was investigated in order to determine the influence on hydrogen purity as suggested by other research groups [5].

Long-term employment of the oxygen carrier also gives fundamental insights into the behavior for the proposed system. The applied material in the experimental plant featured a high resistance towards mechanical and thermal stress from cyclic reduction and oxidation in long-term operation throughout 70 cycles and more than four months of uninterrupted high-temperature exposure.

In addition, several trace gas compounds in biogas are prone to significantly affect the product gas purity. To investigate the impact of these compounds, the main impurities in common biogas and gasified biomass feedstocks were identified and tested. Their influence on carbon deposition and hence also on the product gas impurities in hydrogen generation phase was investigated in a fixed-bed system. Model compounds were defined for alkanes, cycloalkanes and alcohols and investigated in a fixed-bed system. Carbon depositions were identified especially from co-feeding alcohols and cyclic compounds at higher relative shares applied in reduction phase, although additional steam was applied to enable full conversion the compounds.

Acknowledgements

Financial support by the Klima- and Energiefonds through the Energy Research Program 2015 is gratefully acknowledged.

[1] V. Hacker, A novel process for stationary hydrogen production: The reformer sponge iron cycle (RESC), J. Power Sources. 118 (2003) 311–314. doi:10.1016/S0378-7753(03)00076-4.

[2] R. Zacharias, S. Visentin, S. Bock, V. Hacker, High-pressure hydrogen production with inherent sequestration of a pure carbon dioxide stream via fixed bed chemical looping, Int. J. Hydrogen Energy. (2019) 1–15. doi:10.1016/j.ijhydene.2019.01.257.

[3] S. Bock, R. Zacharias, V. Hacker, High purity hydrogen production with a 10kWthRESC prototype system, Energy Convers. Manag. 172 (2018) 418–427. doi:10.1016/j.enconman.2018.07.020.

[4] S. Bock, R. Zacharias, V. Hacker, Experimental study on pure hydrogen production from biogas in a combined fixed-bed reformer and chemical looping system, in review (2019).

[5] J. Lachén, P. Durán, J.A. Peña, J. Herguido, High purity hydrogen from coupled dry reforming and steam iron process with cobalt ferrites as oxygen carrier: Process improvement with the addition of NiAl2O4catalyst, Catal. Today. 296 (2017) 163–169. doi:10.1016/j.cattod.2017.04.046.

572284 Perovskite Oxygen Sorbents for "Low Temperature" Thermochemical Air Separation: Correlating Compositions with Redox Performance

As a potentially more efficient alternative to cryogenic air separation, thermochemical air separation via cyclic redox reactions of an oxide-based oxygen sorbent has been extensively investigated recently. Although a number of promising oxygen sorbents have been proposed and investigated, further improvements in sorbent performance through fundamental understanding of the relationships between oxygen sorbents' structural/compositional properties and redox performance are highly desirable.

In this study, we systematically investigated the effects of A and B site dopants on the oxygen uptake/release properties of $SrFeO_3$ family of perovskites as oxygen sorbents. Addition of dopants to the A site, namely Ba and La, or the B site, Cu or Mn, increased the oxygen capacity for temperature and pressure swings when compared to undoped $SrFeO_3$. Based on these results, the Mn sample underwent 1000 redox cycles where the sorbent was subjected to oxygen partial pressure swings between 20% O_2 and inert at 600 °C. The Mn doped sample proved to be stable in terms of oxygen storage capacity, with less than a 3% decrease in capacity. The redox kinetics were similar for the last few cycles when compared to the first few cycles. DFT calculations, in conjunction with the experimental results, indicated that oxygen vacancy formation energies are correlated to initial decomposition temperature for the samples investigated. However, the oxygen storage capacities of the doped samples did not show good correlation with the vacancy formation energies. It was further determined that the dynamic nature of the redox process makes it difficult to use a single vacancy formation energy as the descriptor for the oxygen storage capacities of the sorbents. Instead, a systematic approach was developed to correlate the oxygen storage capacities with the sorbents' compositional properties and vacancy formation energies. The findings from this study can potentially be used to develop improved sorbents for thermochemical air separation.

567742 MoVNbTeO for Chemical Looping - Oxidative Dehydrogenation of Ethane

Chemical looping (CL) is a concept that depends on conveying lattice oxygen, instead of gaseous oxygen, to a reaction by means of a metal oxide, termed, in this context, an oxygen carrier. This enables physical or temporal separation between the oxidation and reduction steps. Oxidation occurs when the oxygen carrier is contacted with air in the first reactor, while reduction takes place in the second reactor by reaction with a reducing agent. CL might be employed for the production of chemicals via selective oxidation, or oxidative dehydrogenation (ODH), reactions, which can render these reactions safer, more selective and more intensive. In addition, CL would result in reducing CO_2 and NO_x emissions. However, the key challenges to adopt CL for selective oxidations include quantifying the benefits and drawbacks of applying the concept, in addition to the development of a stable redox catalyst.

The present study investigates the potential of chemical looping ODH (CL-ODH) of ethane. A multimetallic catalyst, namely MoVNbTeO, was chosen as the primary catalyst for this study, given its high activity for ODH. The choice was confirmed by a wide catalyst screening, spanning other active metals and oxygen carriers. The activity of MoVNbTeO was optimised for CL by tuning preparation conditions, most importantly calcination regimes. XRD, TGA and SEM-EDS confirmed the improved crystal structure, high oxygen capacity and homogenous distribution of metals for the developed catalysts.

To further enhance ethylene yield, particulate oxygen carriers composed of bismuth oxides supported on ceria-zirconia were physically mixed with the catalyst. This class of oxygen carriers can selectively combust hydrogen in the presence of ethylene. The results showed that the presence of the oxygen carriers slightly improved conversion and CO_2 selectivity, without affecting overall ethylene yield.

Catalyst testing indicated that the developed catalyst can realise 50% ethylene yield, with 90% selectivity and 55% conversion at 450°C. This represents a significant improvement compared to CL-ODH catalysts described in the literature that can achieve similar yields only at temperatures as high as 850°C.

576625 Improving Fe_2O_3 Reactivity with CO and CH_4 in Chemical Looping Systems By Low Percentage 1st-Row Transition Metal Dopants: A Theoretical and Experimental Work

Chemical looping (CL) technology, which includes chemical looping combustion (CLC) and gasification (CLG), is one of the most promising platform in the field of clean energy due to its capability in simplified inherent CO_2 capture and value-added chemical production. One crucial factor in dictating the success of a CL process is the reactivity of oxygen carriers. In the CL technology developed at The Ohio State University (OSU), iron oxide is employed as the active oxygen carrier due to its abundance and low cost. In this work, we focus on improving the reactivity of Fe_2O_3 with CO and CH_4 by using low percentage (1%) 1st-row transition metal (TM) dopants. A systematic dopant screening strategy is developed via the combination of *ab initio* density functional theory (DFT) simulations and experimental validations conducted in TPR (temperature programmed reduction). Reaction pathways of CO and CH_4 oxidation on the most thermodynamically stable Fe_2O_3 (0001) surface with and without dopants and defects were investigated, and the rate-limiting steps (RLS) were identified. Descriptors that correlate the energy barrier of the RLS with different dopants were found. Our results demonstrate that copper is the most efficient dopant among 1st-row TMs as it can lower the reduction initiation temperature with CO and CH_4 by 50 and 150 °C, respectively. Such a decrease in operating temperature can expand the operational matrix for the autothermal operation, and facilitate the development of CL processes.

557230 Unconventional Oil and Natural Gas - Science & Technology Advancement

Despite the significant growth in the production of Unconventional Oil and Gas (UOG) resources during the shale revolution, very little is known in detail about recovery efficiency in these hydraulically-fractured reservoirs. The limited information that is available suggests that recovery factors (the ratio of produced resource to total in-place resource) are typically quite low: approximately 20 percent in gas-rich shale reservoirs and less than 10 percent in liquid-rich plays. By increasing the recovery factors of hydrocarbons, the environmental impact of these activities would also be reduced.

Therefore, the primary research challenges include, but are not limited to: 1) understanding the nano-scale structure of UOG reservoirs in order to more effectively create and control stimulated rock volume; 2) improving recovery efficiency to reduce the incidence of stranded/unproduced resources and the related lost economic and energy security benefits; 3) understanding the anticipated continued growth in recoverable resource volumes with technology advance; and 4) accurately assessing the potential nature, scale, and rate (intensity) of future development in various areas of the country to guide assessment of region-specific and environmental impacts.

Hydraulic and natural fracture diagnostic technologies have not advanced beyond microseismic monitoring techniques, and there is a need to improve the characterization of the far-field fracture geometry for recovery factor calculations and improvement. Advanced diagnostic and monitoring tools, as well as subsurface models, can be capable of predicting induced seismic events and may provide insight and benefit to our understanding and modeling of the stimulated reservoir volume during hydraulic fracturing operations.

Once the geometries of natural and induced fractures are better understood, new physical and chemical treatments can be formulated for each specific well to achieve maximum recovery factors. Advanced stimulation technologies, such as energized foams or other novel fluids could work in tandem with advanced fracture diagnostic tools to improve recovery efficiency and reduce potential environmental impacts.

575806 The Potential for Enhanced OIL Recovery in the Bakken Petroleum System

The Williston Basin's Bakken petroleum system, which includes the Bakken and Three Forks Formations, is a world-class unconventional tight oil play with original oil-in-place (OOIP) estimates ranging from 300 billion to 600 billion barrels. Matrix permeability in the Bakken is typically on the order of micro- to nano-Darcies, and hydraulic fracturing techniques are necessary to produce oil from the reservoir. Despite the enormous resource, the U.S. Geological Survey estimates the technically recoverable reserves for the Bakken are only 7.4 billion barrels, with primary recovery factors for individual wells typically ranging from 4% to 10%. The massive size of the resource combined with low recovery factors means business as usual will leave an enormous amount of oil behind, and even small improvements in productivity would yield billions of barrels of incremental oil. The Williston Basin also holds world-class lignite coal reserves. Several large lignite coal-fired power plants in North Dakota and Saskatchewan operate within 100 km of the most oil-productive areas of the Bakken Formation. The juxtaposition of a need to improve the productivity of a world-class oil resource with a desire to manage carbon dioxide (CO_2) emissions from nearby power plants has led to an interest in the potential to use CO_2 for enhanced oil recovery (EOR) and associated storage in the Bakken Formation. There is also an interest in using hydrocarbon gases (e.g., methane, ethane, and propane, also referred to as "rich gas"), the coproduction of which may inhibit oil production because of inadequate gathering and processing infrastructure and tight regulations on flaring. Although flaring associated with Bakken oil production has been reduced significantly in recent years, as of January 2019, approximately 507 million cubic feet per day (MMcfd) of rich gas was being flared, accounting for 18% of the rich gas produced in association with Bakken oil production. The North Dakota Industrial Commission, which regulates the oil and gas industry in North Dakota, has a stated goal of reducing the amount of flared rich gas to 10% or less of the total associated gas produced in the state by November 2020. Therefore, reduction of emissions due to flaring from the Bakken is a high priority for government and industry stakeholders in North Dakota.

From 2012 to 2018, the Energy & Environmental Research Center (EERC) conducted a series of consortium-driven research programs focused on potential approaches to EOR in the Bakken petroleum system. The U.S. Department of Energy (DOE) and industry consortium-funded projects used new and existing reservoir characterization, laboratory experimental data, state-of-the-art modeling, and two field-based injection tests to examine the viability of EOR using CO_2 or rich gas. The lab- and modeling-based studies demonstrate that CO_2 and rich gas can permeate the Bakken and Three Forks matrix, largely through diffusion, to mobilize oil. Those studies also showed that CO_2 will preferentially mobilize lower-molecular-weight hydrocarbons. The modeling studies indicated that huff 'n' puff EOR schemes may be the most effective approach to EOR in the Bakken petroleum system. The laboratory and modeling activities were used to inform the design and execution of two pilot EOR tests, one using CO_2 and the other using rich gas as the injection fluid.

In 2017, XTO Energy in partnership with the EERC conducted a CO_2 injection test in a vertical well completed in a nonshale, but still unconventionally tight and productive, zone of the Bakken Formation referred to as the Middle Bakken. The objectives of the test were to determine the injectivity of an unstimulated Middle Bakken reservoir (i.e., a reservoir that had not been hydraulically fractured) and the ability of injected CO_2 to permeate the matrix and mobilize oil. The well completion program did not include the use of hydraulic fracturing and proppant. Upon perforation, the well did not flow to surface, but oil samples were collected from the well before injection. Approximately 99 tons of CO_2 was injected over 4 days. The CO_2 was allowed to soak for 15 days. Reservoir pressure and temperature were monitored during all stages of the test using downhole gauges. During the flowback period, gas composition was monitored, and fluid samples were collected. Preinjection and postinjection oil samples were analyzed for oil composition to determine the molecular weight distribution of the hydrocarbons. Pulsed-neutron logs were also run before and after injection to evaluate the vertical distribution of the CO_2

in the near-wellbore environment.

Injectivity of the unstimulated Middle Bakken was found to be low, with stable CO_2 injection rates between 6 and 12 gallons per minute and bottomhole pressure during continuous injection from 9400 to 9470 psi. The CO_2 penetration radius was calculated to be 50 to 70 ft. However, numerical simulation indicated it may be as much as 140 ft. During flowback, oil flowed to the surface briefly, during which time fluid and gas samples were collected. Analyses of the preinjection and postinjection oil showed that the composition of the postinjection oil samples had greater amounts of lower-molecular-weight hydrocarbons than the pretest oils. Interpretation of the results from the field test suggest that although matrix injectivity is low, injected CO_2 can penetrate the Middle Bakken and mobilize oil from the matrix.

The data from the CO_2 field test were used in simulation modeling exercises to gain further understanding of the flow characteristics of CO_2 in unconventional tight oil formations using a variety of EOR schemes. The simulations indicated that the cyclic multiwell huff 'n' puff (CMWHP) approach showed the best performance in terms of EOR. In the best cases, the CMWHP scheme was predicted to more than double the oil recovery factor of a well. The CMWHP scheme would include a sequence of alternating injection, shut-in (soak), and production using three or more wells. In 2018 Liberty Resources, in partnership with the EERC, initiated a pilot-scale EOR test using rich gas as the injection fluid and a CMWHP scheme for injection and production. The hypothesis being tested by the rich gas pilot testing activities is that CHWHP can be used to maintain conformance of injected fluids within the target reservoir, improve oil recovery response, and utilize rich gas components that might otherwise be flared. Rich gas injection started in the summer of 2018, and as of the end of March 2019, over 76 MMscf of gas had been injected sequentially into four wells during six different injection periods. Operation of the pilot is expected to continue into the summer of 2019.

The results from the laboratory-, modeling-, and field-based studies indicate there is a tremendous potential to conduct EOR operations in the Bakken petroleum system. The results of the CO_2 lab studies and field test were integrated into simulations of various EOR scenarios. Those simulations showed incremental oil recovery from the injection of CO_2 into a Bakken or Three Forks reservoir as high as 5.4% of OOIP. With OOIP estimates ranging from 300 billion to 600 billion barrels, the modeling results, therefore, suggest that the use of CO_2 for EOR in the Bakken petroleum system may yield between 16 billion and 49 billion barrels of incremental oil. The laboratory-based studies show that CO_2 and rich gas behave similarly in terms of their ability to permeate tight Bakken rocks and mobilize oil. While the rich gas EOR studies and pilot test are still ongoing and results from those activities will not be available until 2020, it is anticipated that the potential impacts of using rich gas for EOR in the Bakken will be of a similar magnitude.

559689 Gas Transport in Shale Matrix Coupling Multilayer Adsorption and Pore Confinement

Accurate prediction of gas transport in shale formations mainly consisting of micro- and nano-scale pores is a great challenge since most existing models only consider monolayer adsorption and the underlying multi-physics are either partially or completely overlooked. In this work, we propose a comprehensive gas transport model integrating multilayer adsorption, surface diffusion, real gas effect, and pore confinement effect. The new analytical model is developed based on Bravo layer-sequence-model and then upscaled using an Effective Medium Approximation (EMA) method, where the generalized Brunauer-Emmett-Teller (BET) model, the modified Peng-Robinson equation of state (EoS) model, and the Sutton's viscosity model are innovatively incorporated. The newly developed model has been successfully validated against simulation and experimental results with the assistance of an efficient iterative ensemble smoother (ES) algorithm. It is observed that the pore confinement effect is of importance when the pore size is smaller than 50 nm. The increase of gas viscosity due to the real gas effect is significant under reservoir conditions. The apparent permeability is found to increase greatly as the adsorption layer number increases, implying that the application of Langmuir model in existing gas transport models may substantially underestimate it. Given inorganic nanopores, the viscous flow is important at any pore sizes and the Knudsen diffusion plays a role in the pores that are less than 5 nm under reservoir conditions. As for organic nanopores, the contribution of surface diffusion is tangible when the pore size is below 150 nm and the Knudsen diffusion is negligible under high pressures.

565990 Management of Iron Precipitation As a Solution for Decreased Petroleum Production from the Utica/Point Pleasant Unconventional Play

The development of U.S. unconventional resources has had a dramatic impact on global energy supply, with the U.S. becoming the world's leading producer of petroleum and natural gas. This change has been beneficial to the public lowering energy prices, while increasing energy security. Unconventional (or shale) resources differ from conventional reservoirs in a multitude of ways. Production wise unconventional plays see significantly larger production drops, about 75% [1], after one year compared to 5% in conventional [2]. Unconventional wells also differ structurally and geologically as they consist of micro-pore and mainly nano-pore structures of interbedded carbonate and hydrocarbon containing shale layers which serve as the source for reactions with fracturing fluid. A majority of the reactions that occur during fracture process happen in the early stages, as the fracturing fluid has yet to come to equilibrium with the reservoir. While some reactions will create the pathways for hydrocarbons to flow through the well; others may cause the shale surface to be coated with insoluble barium, calcium, strontium or iron based solids/precipitate. These insoluble solids tend to clog the fracture and pore networks creating large decreases in production. Prevention of scaling is vital for maintaining well integrity, which is usually achieved by adding scale inhibitors to the fracturing fluid. However in some wells, a scale inhibitor may be incompatible with well geology or ineffective once equilibrium has been reached resulting in scaling and decrease production. Understanding, the mechanism of shale-fluid interactions helps is identifying the effective scale inhibitor or modifying the fracturing fluid best better performance of the well.

Ohio University (OHIO) is investigating scale formation as a means for the large decrease in production seen in unconventional wells, with a particular focus on the Utica/Point Pleasant (UPP) shale play. In this work, shale-fluid interactions are investigated using experimental trials in a batch reactor at reservoir conditions. Previously, it was found that iron based precipitation are forming in UPP wells under experimental trials. This study will evaluate modifications to hydraulic fracturing formula chemistry in order to evaluate the effect on iron precipitation. Pre- and post-trial analysis of the shale surfaces will be performed using Scanning Electron Microscopy (SEM) paired with Energy-dispersive X-Ray spectroscopy (EDS) and Laser-Induced Breakdown Spectroscopy (LIBS), and fluids will be analyzed using Inductively Coupled Plasma Optical Emission Spectrometry (ICP-OES) and Ultra-Violet Visible spectroscopy (UV-Vis). The effect of various commercially available inhibitors and their concentrations will be evaluated. In addition, modeling of the shale-fluid interactions using Geochemist Workbench will be performed and presented.

[1] H. Wachtmeister, L. Lund, K. Aleklett, and M. Höök, "Production Decline Curves of Tight Oil Wells in Eagle Ford Shale," *Nat. Resour. Res.*, vol. 26, no. 3, pp. 365–377, Jul. 2017.

[2] M. Höök, R. Hirsch, and K. Aleklett, "Giant oil field decline rates and their influence on world oil production," *China Energy Effic.*, vol. 37, no. 6, pp. 2262–2272, Jun. 2009.

566685 Reactive Transport in Shale Matrix after Fracturing Fluid Imbibition

The porosity and permeability of shale matrices are critical for hydrocarbon transport, and thus directly impact production efficiency. After hydraulic fracturing fluid is injected, it is imbibed by shale and cause a series of chemical reactions, including dissolution of primary minerals that increases porosity, and precipitation of scale minerals that occludes porosity, both impacting hydrocarbon flows through the matrix. Our previous experimental work shows that these chemical reactions extend from the fracture into the matrix by micrometers to centimeters, varying with respect to shale mineralogy, fluid composition, and reaction types. Based on our experimental findings, we built a reactive transport model as part of this study. After successfully matching the modeling results with the experimental observations, we obtained the chemical reaction network governing shale-fluid interactions, which can be readily incorporated into shale transport models to explore the effects of chemical reactions. Using our model, we further explored the influence of various factors on chemical alteration. The modeling results strikingly show the difference between domains with and without continuous injection of fracturing fluid With continuous fluid flow in the main fracture, barite ($BaSO_4$) scale is a high concern because it significantly reduces porosity. Iron-bearing scale formation is negligible because the high fluid-to-solid ratio dilutes the iron concentration. In contrast, with less fluid flow, the shale matrix experiences less barite scale formation, but iron concentration can increase locally resulting in iron(III)-hydroxide precipitation in the matrix as well as in the fluid phase occupying pore space in the fractures/microcracks. The fundamental understanding of reactive transport processes gained from this study helps us to evaluate chemical damage in unconventional reservoirs due to fluid imbibition. The numerical modeling framework will be used to explore basin-specific scale-control strategies.

574655 Surfactant- or Nanoparticle-Stabilized CO_2-in-Oil and Natural Gas-in-Oil Emulsions and Foams for Completely Waterless Hydraulicfracturing

One of the most promising, completely waterless hydraulic fracturing fluids that reached commercialization during this decade was gelled liquid propane and gelled LPG. The thickening was accomplished via the introduction of carefully designed LPG-soluble phosphate esters tailored to remain somewhat soluble in the LPG upon the dispersion of an aqueous solution with trivalent metal ions that cross-linked the phosphate esters. In an attempt to generate completely waterless hydraulic fracturing fluids based on fluids other than propane, we began by assessing these phosphate-ester-based formulations in liquid ethane and liquid CO_2. In this presentation, we will demonstrate that cross-linked phosphate esters are indeed incredibly effective in propane, however when the aqueous crosslinking solution was added to single phase ethane-phosphate ester or CO_2-phosphate esters solutions, the cross-linked complex immediately precipitated and no thickening resulted.

As a result, we turned our attention to combinations of CO_2 and mineral oil (oils were selected that would not become completely miscible with CO_2 at high pressure). CO_2-energized mineral oil (this term refers to a CO_2-oil mixture at high pressure in which a substantial amount of CO_2 dissolves in the mineral oil) has been commercially applied for water-sensitive formations during this decade, but these mixtures contain relatively low proportions of CO_2 (e.g. 30wt%). In an attempt to increase the proportion of CO_2, we successfully designed non-fluorous, mineral oil-soluble, silicone oil-alkyl copolymer surfactants capable of stabilizing high apparent viscosity CO_2-in-oil (C/O) emulsions up to ~90 vol% liquid CO_2.

In this study, we demonstrate that one can also stabilize these (C/O) emulsions with non-fluorous, surface-functionalized, mineral oil-dispersible nanoparticles (NP). NP-stabilized (C/O) emulsions with apparent viscosities up to 14 cP can be generated with surface-modified nanoparticles despite a low driving force for the nanoparticles to adsorb to the interface, given the low interfacial tension. We systemically designed the nanoparticles to be soluble in mineral oil and interfacially active at the CO_2-oil interface through the nature of the surface modification. We initial grafted a C_{16} alkane, a lipophilic ligand, to the surface to make the nanoparticles soluble in mineral oil and then grafted a secondary non-fluorous CO_2-philic ligand to make the nanoparticle interfacially active. The nanoparticles were readily soluble in mineral oil and lowered the CO_2-oil interfacial tension from 4.7 mN/m to as low as 3.1 mN/m. Furthermore, the most interfacially active nanoparticles were able to stabilize a C/O emulsion at shear rate as low as 370 s^{-1} and the emulsion droplet size decreased to 10 mm with increasing shear rate. The droplet size and apparent viscosity are compared for various nanoparticle surface modifications. The emulsion apparent viscosity generally increased with an increase in the mineral oil phase viscosity. The emulsions were stable for at least 72 hrs, a long enough time frame for application as a fracturing fluid to suspend proppant. The emulsions formed with nanoparticles had comparable viscosities and stabilities relative to those formed with our previously designed silicone oil-alkyl copolymer surfactants. The mechanism of emulsion stability is compared for the polymer and nanoparticle stabilized emulsions.

Finally, we will present out initial findings for another type of completely waterless hydraulic fracturing fluid; natural gas-based fluids. Since methane is a far weaker solvent than ethane or CO_2, phosphate esters cannot be used to thicken high pressure natural gas. Therefore, we turned our attention to natural gas-in-mineral oil foams. The benefits of this waterless fluid include the ability to acquire the natural gas from the same field's production and field storage as LNG, the ability for the flowback natural gas to simply blend with the natural gas production, and the ability of the flowback mineral oil to simply blend with the produced hydrocarbon liquid products (condensate and/or crude oil). This minimizes trucking requirements and water-related processing and eliminates the need to vent gases. Further (unlike the cross-linked phosphate ester components), the novel non-fluorous, mineral oil-soluble surfactants that we have designed or selected for stabilizing these natural gas-in-mineral oil (NG/O) foams will not cause downstream processing problems in refineries. Using methane as a model component for a lean natural gas, we have shown that it is possible to generate NG/O foams with high apparent viscosity, although their stability (to date) is not as great as that exhibited by the C/O emulsions.

570863 Boiling Points of Hydrocarbons in Nanoporous Media at High Pressures Using Differential Scanning Calorimetry Measurements

The United States produced close to 11 million barrels of oil per day in the first three months in 2019. A significant portion of this oil comes from tight oil plays in the Permian basin and other shale plays. Imaging and characterization studies have shown that the pore sizes in rocks from tight oil plays are of the order of 5-10 nanometers. It is known that fluid properties are different in such small pores. The purpose of this paper is to measure boiling points of hydrocarbons at pressures higher than atmospheric in materials that consist of pores of similar sizes and compare them to bulk boiling points.

Two silica based mesoporous materials with internal average pore size of about 4 nm were first synthesized. SBA-15 has ordered hexagonal pores while SBA-16 is characterized by ordered cubic pores. Boiling points of heptane and toluene were measured in these materials using a high-pressure differential scanning calorimetry instrument. Boiling points were also measured in the bulk and with sand (with no internal pore structure) for comparison. Boiling points were measured at 1, 7 and 13 bar pressures.

The boiling points were higher at all of the pressures studied, and for all the fluids for SBA-15 and SBA-16 in comparison to the bulk and sand measurements. It was thus possible to generate vapor-pressure curves under confinement for the fluids studied in this paper.

577144 Methane Pyrolysis with a Molten Cu-Bi Alloy Catalyst

Current methods of hydrogen production from methane generate more than 5 kilograms of CO_2 for every kilogram of hydrogen. Methane pyrolysis on conventional solid heterogeneous catalysts produces hydrogen without CO_2 but the carbon coproduct poisons the catalyst. This can be avoided by using a molten metal alloy catalyst. We present here a study of methane pyrolysis on mixtures of molten Cu-Bi alloys as the catalyst. We find that molten Cu-Bi is an active catalyst, even though pure molten Bi or Cu are not. Surface tension measurements and constant temperature *ab initio* molecular dynamics simulations indicate that the surface is enriched in Bi and that the catalytic activity is correlated with the concentration of Bi at the surface. Bader charge analysis indicates that bismuth donates charge to copper and the most stable configurations of dissociated methane on these liquid surfaces are those with CH_3 coordinated to the bismuth surface atoms and H coordinated to Cu. Methane-activation energy barriers were calculated using the nudged elastic band (NEB) method to be 2.6 eV and 2.5 eV, on different sites on the $Cu_{45}Bi_{55}$ alloy. The computed barriers are in agreement with the experimentally measured apparent activation energy of 2.2 ± 0.2 eV.

576031 Exploring Thermodynamic and Efficiency Trade-Offs in a High-Pressure Chemical Looping Hydrogen Production System

Iron-based Chemical looping is a promising technology that can reduce costs from SMR for H2 production. The chemical looping process uses three reactors, which eliminate the need for multiple processing units in the SMR process, and increases H_2 production. Previous studies have shown promising results in quantifying the thermodynamic limits of the system at atmospheric pressures[1]. Additionally, most of the design philosophy in a chemical looping system follows the principle of minimization of solids circulation rate following the circulating fluidized bed derived heuristics. This study seeks to quantify the feasibility and tradeoffs associated with high-pressure chemical looping applications while relaxing the assumptions for heuristically inspired minimum solids flowrate design. An initial isothermal quantification using the Gibbs free energy minimization approach and supplanted by kinetic assumptions will be used to develop a base-case performance for the chemical looping system. This initial design will be updated with an adiabatic and autothermal operation for the system, wherein temperature swings of solids and pre-heats will be synchronized such that the system has maximum Hydrogen production with minimum energy consumption. Finally, deterministic (SQP-type) and stochastic (simulated-annealing type) optimization solvers will be used to relax the assumption of minimum solids flowrate in-order to increase Hydrogen production efficiency. The preliminary results of this study provide an initial thermodynamic quantification of high-pressure chemical looping solids operation and explore tradeoffs between compressing H_2 versus operation of a solids circulation loop at high pressures. Specific insights into the delicate interplay between Hydrogen production efficiency, solids circulation fluxes and system design temperature provide an essential understanding and motivation for specific designs as well as scaled operating condition recommendations. Additionally, Individual and synergistic effects of variables like the number of compressor and expander stages, variation in reactant space hourly velocities and the trade-offs associated with using pinch and transshipment type technology for heat exchanger network synthesis are also investigated. Preliminary results show that operating chemical looping system at operating conditions that deviate from minimum solids circulation rate can provide equivalent or even higher H_2 production efficiencies than those obtained from following conventional heuristics.

1. Kathe, Mandar V., et al. "Hydrogen production from natural gas using an iron-based chemical looping technology: Thermodynamic simulations and process system analysis." *Applied energy* 165 (2016): 183-201.

573005 Development of a High-Purity Hydrogen Production Skid Unit for Hydrogen Refueling Stations

A high-purity hydrogen (99.999 %) production skid unit (200 kg hydrogen/day, called as *KIER HyPU-100*), comprising natural gas-steam reforming, water-gas shift reactors and vacuum pressure-swing adsorption (VPSA) process, for hydrogen refueling stations has been developed. The system should satisfy the main prerequisites for actual application such as 1) higher than 80 % hydrogen production efficiency on the basis of higher heating values (HHV) and 2) lower than 1 ppm of CO concentration at the final outlet of the hydrogen refining process. In order to guarantee the high efficiency, the heat exchanging network in the reforming process was optimized by the commercial process simulation software package and the heat and mass transport phenomena in the reformer were also intensively studied by the computational fluid dynamic models. In addition, the optimal VPSA process has also been developed to purify the hydrogen from the product gas of the water-gas shift reactor. The performance of the VPSA is not only related to the hydrogen purity but also to the hydrogen production efficiency since the recovery ratio of high-purity hydrogen to the total reformate affects the amount of both product hydrogen and make-up natural gas fed into the burner of the reformer. The developed system showed consistently reliable performance within the long-term test sessions (hydrogen production efficiency: $> 80\ \%_{HHV}$, hydrogen purity: 99.999 %) and the operating data was compared with the mathematically predicted values to prove the feasibility of the design concepts.

564892 Hydrogen Production from Methane Cracking in a Fluidized Bed Process

Hydrogen is widely considered as a major future carbon-free energy carrier that can be utilized for power generation, heating and as a transportation fuel. However, hydrogen is currently produced almost exclusively from fossil hydrocarbons, with the dominant production route being the steam-methane reforming (SMR) of natural gas.

$CH_4 + H_2O \rightarrow CO + 3H_2$ ΔH_r=206 kJ (1)

This process inevitably produces CO_2 emissions to the atmosphere and will thus contribute to global warming. Alternatively, hydrogen can be produced by water/steam electrolysis driven by renewable energy; however this process is currently still burdened with high capital costs and not deployed at a large-enough scale. As a bridging technology hydrogen could be produced from methane via cracking on a suitable substrate.

$CH_4 \rightarrow C + 2H_2$ ΔH_r=74.6 kJ (2)

This process binds carbon in solid form and thus does not directly release CO_2 to the atmosphere. This advantage could potentially outweigh the drawback of not being able to utilize the entire calorific value of methane (c.f. LHV(CH_4)= 802 kJ and 2*LHV(H_2)=488 kJ). Here, we explore the cracking of methane into solid carbon and hydrogen in a laboratory-scale fluidized bed reactor over different supported metal substrates. Produced gases were analyzed by on-line ND-IR measurement and materials characterized by X-ray diffraction and SEM imaging at different stages of the process. For each substrate material, we discuss gas conversions, selectivity, obtained carbon structures and strategies for regeneration of / carbon recovery from the substrate. Based on the results, the on-going development of a chemical-looping based approach for this process will be presented.

575126 Improved Design of a Diesel-Driven Solid Oxide Fuel Cell System for Auxiliary Power Unit Applications

In this study, a diesel-driven solid oxide fuel cell system has been developed for auxiliary power unit applications. Among various hydrocarbon fuels, diesel fuel has many advantages such as high energy density and versatility, and it is often used for heavy-duty applications and remote areas. Diesel itself can be used to generate electricity with a diesel generator, but it produces high pollution and noise, with low energy efficiency. In another way, a diesel autothermal reformer can be introduced to generate hydrogen-rich reformate that can be used for a solid oxide fuel cell stack to generate electricity. With the diesel reformer, the diesel fuel inside a fuel tank of the heavy-duty truck can be directly used to generate hydrogen for the fuel cell system.

With the integration of the diesel autothermal reformer and the solid oxide fuel cell stack, a fuel cell system for the auxiliary power unit application has been developed. For the heavy-duty truck applications, the power output target of the auxiliary power unit was set to one kilowatt to satisfy the electrical needs. Structured catalysts were introduced for the diesel reformer to ensure the mechanical strength of the component. During the development of the diesel reformer, a long-term test for the catalyst component was conducted to ensure its catalytic stability. 2,000 hours of operation was achieved with the structured catalyst for the autothermal reforming.

In commercial diesel fuel, a small amount of sulfur content exists. To protect the fuel cell stack, desulfurization with ZnO based adsorbent was followed by the autothermal reforming process. The temperature of the reformate has to be controlled within the range of the operating window of the desulfurizer to reduce the sulfur content to ppm level. A heat network analysis for the reformer module including the desulfurization process was done and required specifications for heat exchangers were obtained.

Two versions of the integrated fuel cell system were designed with different assumptions for the thermal stability of the solid oxide fuel cell stack. The solid oxide fuel cell stack needs a high operating temperature around 700 deg C, and the thermal stability of the stack determines overall system start-up time. The first version of the system was designed with meeting the requirement of the slow heat-up of the stack. The temperature of the fuel cell stack is increased with the heat from the reformate oxidation in the first design, so it takes some time to reach the operating temperature. However, the start-up time of the fuel cell system can be shortened if the fuel cell stack has enough thermal stability that can sustain the fast heat-up process. Therefore, the system configuration can also be modified at the improved version by using external heat sources during the start-up process. In result, shorter start-up time and simplified start-up process were obtained with the enhanced design of the fuel cell system.

574141 Longevity Study of Microscale Non-Thermal Plasma Reforming of Methane to Higher Hydrocarbons and Syngas

Using a non-thermal electrical plasma, reactions that usually require high temperatures and pressures can take place with reactant chemical species near ambient conditions. Instead of thermal molecular collisions causing reactions, a plasma-reaction system utilizes high energy electrons colliding with reactant species in order to disrupt bonds and ultimately force reactions.

Performing these reactions at the microscale level has the advantage of decreasing the space between the electrodes that create the electrical plasma. By reducing the size, less voltage is required in order to generate these plasmas which reduces the costs of the necessary power electronics. This also decreases the bypass of the system, due to the small volume of the channel compared to the volume of the discharge. The electron temperature can range from several 1,000 to 30,000 °C – several orders of magnitude larger than what would be required by most reactions.

In this study, carbon-hydrogen bonds are broken which allows for longer hydrocarbons to form. By creating longer hydrocarbons from methane, this process transforms electrical energy into chemical potential energy. The major products of this process are syngas, C_2 hydrocarbons and trace amounts of higher hydrocarbons. Wax products have been found to exist downstream from the discharge on the reactor channel walls. One major drawback of this approach is the rapid coking of the reactor channel at long (1+ hours) run times, leading to eventual flow anomalies, unfavorable product diffusion and eventually reactor failure.

Recent results showing how longevity (20+ hour run time) in these systems can be achieved and maintained will be presented.

565371 Reducing Carbon Emissions through Novel Hydrogen Sulfide Processing Strategies to Produce Hydrogen Gas

We present a novel hydrogen sulfide processing scheme that leverages the oxidation of H_2S to produce HI from H_2S, H_2O and I_2. We further propose that the HI generated by this reaction could be decomposed into H_2 and I_2, net producing H_2 from H_2S and H_2O. The reactions comprising this cycle are listed below.

$H_2S + 2H_2O + 3\ I_2 \rightarrow 6\ HI + SO_2$

$6\ HI \rightarrow 3H_2 + 3\ I_2$

With separation steps removing the SO_2 after the first reaction and H_2 after the second reaction, we believe that this chemistry could present an alternative to the Claus Process, with the added benefit of H_2 production.

The development of this invention is motivated principally by the possibly of reducing CO_2 emissions by avoiding H_2 production from natural gas. As nations adopt increasingly strict and widespread sulfur emission standards, the demand for hydrogen gas as a reactant in hydrodesulfurization processes rises. This demand is almost entirely met with hydrocarbon reforming, producing hundreds of millions of metric tons of CO_2. The proposed cycle eliminates the need to generate additional CO_2 for hydrodesulfurization by producing H_2 from H_2S and H_2O. While limited by the amount of H_2S available, even at current levels, the potential for CO_2 emission reductions is significant. Given global trends in desulfurization practices and upcoming governmental actions we anticipate the impact of such a process to continue to rise.

This work includes experiments detailing the first reaction step and process modeling describing the proposed process. In addition, the potential impact on world CO_2 emissions is discussed. Future directions and ongoing experiments illustrate where we intend to take this technology.

579106 Hydrogen Injection in Natural Gas Pipelines for Decarbonization of Power Sector in New York State

Recent advancements in techniques to produce renewable electricity or gases have accelerated the need for new flexible means to store and distribute renewable energy. The availability of renewable electricity from sources such as solar and off-shore wind, is being promoted in New York, providing an opportunity to store energy from these intermittent sources as "hydrogen" and New York's existing natural gas distribution system as a means of delivering hydrogen to intended destinations.

While the Power-to-Gas (P2G) concept is gaining attention, the transport of produced hydrogen is also being discussed and several pilot studies underway, though mainly in the UK, Australia and Japan. We recently initiated an assessment of the current ability of each component of the utility gas distribution system to transmit hydrogen or natural gas/hydrogen blends and identify the requirements for increased use of hydrogen to deliver green power to New York State customers.

We believe that an engineering assessment, safety protocols to handle hydrogen/natural gas blends, understanding changes in key parameters by flowing blends on pipeline infrastructure, including quantification of pipeline life prediction would be necessary for hydrogen management in natural gas pipelines. Together with known data, we calculate that upon achieving total decarbonization of natural gas in New York State can achieve a reduction of 24.89 million metric tons of CO_2 that amounts to 39.0% in the power sector or 15% of New York's total CO_2 emissions.

579216 Hydrogen Production System Using Soec - Evaluation of the Energy By Long-Term Operation

To establish the hydrogen production technology by renewable energy, we are developing hydrogen production system by high-temperature steam electrolysis using solid oxide electrolysis cells (SOEC). With this method, it is possible to produce hydrogen at low electrolytic voltage, as compared with alkaline-water electrolysis or polyelectrolyte film water electrolysis.

For the mass production of hydrogen by SOEC, it is necessary to connect multi-typed cell-stacks and to increase the electrolysis capacity. We have already developed the multi-typed cell-stacks pilot test plant for hydrogen production, and demonstrated that we can scale up with maintaining the performance of each stack, and that we can realize large-scale hydrogen production using this method. Furthermore, since the SOEC operating temperature range is from 600 to 800 °C, by providing heat exchangers in this system, the consumption energy of the whole system can be reduced.

However, considering actual usage for a long period of time, the degradation of the stack could cause the increase of the energy consumption of the whole system. Since the SOEC stack generates heat due to degradation, we considered using this heat effectively. Therefore, in order to clarify the temporal change of energy required for hydrogen production, we evaluated the initial degradation by continuous operation of the multi-typed SOEC stacks. In addition, we examined the energy change of the whole system accompanying the deterioration of the stacks.

In the test, the temperature in the module container was set to 650 °C. A constant current was applied to each line for a total of about 1000 hours.

In order to evaluate the characteristics of each cell-stack, the temperature was raised up to 700 °C every several hundred hours to obtain an I-V characteristic.

The initial deterioration characteristics of the stack were evaluated from the transition of the average cell voltage and the stack temperature of each stack.

The stack's I-V characteristics showed the same characteristics as the initial one for about 500 hours.

After 500 h, the stack temperature gradually increased relative to the set temperature (650 °C.) in module container, and the cell voltage also increased gently as compared to the initial cell voltage.

On the other hand, the consumption energy other than the stack decreased the heating energy of the module container due to the heat generation of the stacks. As a result, there was no significant change in the energy consumption of the entire system in this test plant. This demonstrates that even if the characteristic of the stack is degraded, the system performance can be maintained by using heat effectively.

A part of this study is based on the results obtained from a project commissioned by the New Energy and Industrial Technology Development Organization (NEDO) in fiscal 2014 to 2018.

569428 National Energy Technology Laboratory's National Risk Assessment Partnership (NRAP): Efforts to Reduce Risk and Ensure Successful Storage

The U.S. Department of Energy's (DOE) Office of Fossil Energy (FE) and the National Energy Technology Laboratory (NETL) implement the Carbon Storage Program, which supports research and development of technologies to enable safe, cost-effective, and secure containment of carbon dioxide (CO_2) in geologic storage reservoirs. One of the key methods for accomplishing this is to assess and quantify the potential risks associated with the highly complex storage process. The National Risk Assessment Partnership (NRAP), led by NETL, was created to broadly support the Carbon Storage Program's efforts to assess risk at storage sites by integrating expertise and resources from five DOE national laboratories: NETL, Los Alamos National Laboratory, Lawrence Berkley National Laboratory, Lawrence Livermore National Laboratory, and Pacific Northwest National Laboratory. NRAP utilizes expertise from these laboratories to develop, improve, test, and validate tools and methods; perform foundational experiments; and analyze results and data to address risk and the uncertainty inherent in geologic CO_2 storage. NRAP has developed an opensource toolset to quantify and reduce the risk related to the migration of CO_2 or brine from the storage reservoir, and to quantify and reduce the risk of negative geomechanical impacts due to injection of CO_2. Ongoing work includes efforts to characterize and reduce risk and inform decision making throughout the entire life-cycle of a CO_2 storage project with particular emphasis on the characterization and permitting phases. The ultimate goal of this effort is to help quantify and reduce the risk of CO_2 releases to the atmosphere and accelerate the viability and potential widespread deployment of carbon storage technologies. An overview of the results of the NRAP effort to date, the current research and development, and the future direction of the project will be discussed.

576167 A Comparison of the Effectiveness of Three Types of Borehole Seismic Technologies for Monitoring CO_2 Injected into Carbonate Pinnacle Reefs

Battelle manages the Midwest Regional Carbon Sequestration Program (MRCSP) which is a multi-year program funded by the U.S. DOE Regional Carbon Sequestration Program (RCSP) and private partners that is aimed at developing information, technologies and capabilities related to carbon capture and storage (CCS) and carbon capture, utilization and storage (CCUS). The MRCSP program is in its third of three phases, the goal of which is to inject and monitor one million metric tons of carbon dioxide (CO_2) into a series of pinnacle reef carbonate reservoirs in northern Michigan to determine the efficacy of this geologic system for CCS/CCUS. Thousands of pinnacle reef carbonate reservoirs are present in Michigan, most are oil-bearing or gas-bearing. While the individual reefs are relatively small structures, collectively the reefs have an enormous capacity for storing CO_2. The pinnacle reefs are attractive low-risk CO_2 storage reservoirs because they are encased by a thick sequence of low permeability carbonate, salt, and shale that serves to prevent the CO_2 from migrating upward due to its buoyancy. Despite these desirable characteristics, monitoring the position and quantity of the injected CO_2 may be necessary to demonstrate that the CO_2 is remaining within the intended storage formation. Seismic monitoring has been shown to be effective for delineating CO_2 in clastic rocks (sandstone) having high porosity; however, there are few examples where seismic monitoring has been conducted in carbonates. This paper will present results of three types of seismic monitoring, all borehole techniques, that are being deployed in the third phase of MRCSP and discuss challenges associated with each type due to the natural geologic system and other anthropogenic factors such as oil- and gas-well construction practices.

Three types of borehole seismic monitoring were conducted. The objective of each monitoring study was to evaluate the effectiveness of the respective seismic technology for detecting and delineating the horizontal and vertical distribution of (free-phase) CO_2 in a deep carbonate reservoir (approximately 4,000 to 5,000 ft below surface). In one study, time-lapse Vertical Seismic Profile (VSP) seismic surveys were performed in one reef (Reef A), including one baseline VSP in 2013 before starting CO_2 injection and one repeat VSP in 2016 after injecting approximately 200,000 metric tons of CO_2. Time-lapse Distributed Acoustic Sensing (DAS) VSP seismic surveys were performed in a second reef (Reef B), including one baseline DAS VSP before starting CO_2 injection and one DAS VSP after injecting approximately 87,000 metric tons of CO_2. In the third study, a single cross-well seismic survey was implemented in Reef B, also after 87,000 metric tons of CO_2 had been injected, to attempt to discern the horizontal and vertical distribution of CO_2 based on the difference in the response of compressional (P) wave and shear (S) wave velocities. Data acquisition has been completed for all three studies and data processing is either completed or nearly complete.

571039 Monitoring of an Active CO_2-EOR Field Using Charged Wellbore Casing Controlled Source Electromagnetics

Injection of carbon dioxide into depleted oil and gas reservoirs during tertiary recovery (CO_2-EOR) offers numerous tangible benefits including long term storage as well as production of otherwise inaccessible hydrocarbons. In the absence of production activities, brownfield sites are often excellent candidates for CO_2 sequestration due to the extensive characterization that typically has occurred in addition to the presence of infrastructure including legacy boreholes and gas processing facilities. Additionally, hydrocarbon production can help to offset storage costs. For these reasons a substantial component of CO_2 sequestration is anticipated to remain in active and former oilfields.

However, in spite of the benefits, CO_2-EOR operations present unique challenges in order to ensure efficiency and efficacy. Three phase reservoir systems are complicated due to variable phase injection schedules and incompletely characterized fluid-fluid and fluid-matrix interactions. Reservoir models are powerful tools for management, but uncertainty regarding these models can be large and difficult to quantify. Subtle changes in properties including relative permeability and capillarity can have profound impacts on the predicted state of the reservoir at any point in time. This ambiguity in reservoir models can have large impacts with respect to optimal management of the field and risk analysis. Improved methods for subsurface monitoring of these dynamic reservoir systems are therefore needed.

Geophysical methods are a natural means by which to achieve reservoir monitoring and model validation. However, in pressure managed fields such as CO_2-EOR sites, time lapse seismic methods often contain very little signal. Gravity anomalies are often below detection limits. However, electrical methods are typically sensitive to the changes in saturation associated with gas and water injection, and are therefore investigated here. As an added constraint, ideal monitoring technologies need to be minimally invasive and also minimally impactful on oilfield activities.

Most CO_2-EOR sites have been in production for many years and contain numerous legacy boreholes which are not actively in use. These offline legacy boreholes can be used as long deep electrodes which penetrate into the reservoir of interest in controlled source electromagnetic (CSEM) surveys. The use of legacy boreholes as electrodes increases the depth of investigation of charged wellbore casing controlled source electromagnetics (CWC-CSEM) in comparison to conventional CSEM surveys with minimal additional cost. Surface monitoring of magnetic and electric fields are achieved using commodity magnetotelluric (MT) sensors. As such, MT data can be acquired in tandem with CWC-CSEM surveys providing additional information.

As part of an ongoing project funded by the Department of Energy and the National Energy Technology Lab, CWC-CSEM monitoring at an active CO_2-EOR site has been underway for the past several years. Interpretation of the geophysical data is constrained by--and also provides feedback to--reservoir models. As such, the CWC-CSEM method has been shown to provide cost-effective monitoring of tertiary recovery fields. This paper will present findings from the project as well as highlight the approach taken in field acquisition and data processing.

578109 MVA Lessons Learned from Ccus Project at the Farnsworth Unit, Texas

The Southwest Partnership on Carbon Sequestration (SWP) has been conducting long-term Phase III activity at the active Enhanced Oil Recovery (EOR) field in the Anadarko Basin of northern Texas (Farnsworth Unit (FWU)). To date, over 800,000 metric tonnes of anthropogenic CO_2have been successfully stored at the Farnsworth Unit since the project began monitoring the injection (2013). Since 2013, the CO_2-EOR project at the FWU field has been extensively monitored for fluid injection, production and migration, including any fluid migration beyond the reservoir zone. Activities include fluid volume accounting, reservoir and USDW chemical variability, multiple repeat seismic methods, surface CO_2leak detection, and multiphase tracer studies.

Beginning in early 2019, the SWP has shifted to "post-closure" monitoring activities. These MVA activities will be a continuation of critical risk-based activities necessary to evaluate fluid ($CO_2$and/or brine) migration (intended or unintended) at the FWU. The ongoing activities include regular monitoring of USDW groundwater, surface CO_2emissions, time-lapse seismic for CO_2plume assessment, continuous monitoring of microseismicity, and downhole sensors to monitor temperature/pressure. The FWU operator will continue to utilize $CO_2$for EOR activities and the SWP will continue to track fluid injection/production/flare data accounting. The MVA program at the FWU will utilize new or updated activities to provide additional data to other SWP working groups in an attempt to resolve any gaps in understanding of the Farnsworth Unit EOR system.

The MVA program will continue to evaluate past, current and future monitoring activities for efficacy and lessons learned. The best practices of FWU monitoring will serve as a blueprint for regional, economically-viable, commercial-scale CCUS.

Funding for this project is provided by the U.S. Department of Energy under Award No. DE-FC26-05NT42591.

575453 Geologic Carbon Storage of Anthropogenic CO_2 Under the Colorado Plateau in Emery County, Utah

Geologic carbon storage (GCS) is a promising technology for storing large volumes of anthropomorphic CO_2 effectively and permanently. Numerical simulations are an integral part of site selection and characterization for any promising GCS site. In this study, we simulate the injection of CO_2 for geologic storage in the Navajo Sandstone Formation in Emery County, Utah, a promising storage complex for commercial-scale sequestration. Carbon Dioxide is sourced from regional power generation stations and heavy industries throughout Utah, with a particular focus on the emissions from Hunter Power Plant near Castle Dale, Utah.

As part of the CarbonSAFE Rocky Mountains Phase I project a regional GCS analysis was undertaken to understand the efficacy of storing CO_2 emissions from the power generation sector in central Utah's favorable geology. A geological model of the area encompassing the Huntington/Castle Dale/Ferron and San Rafael Swell area was constructed from well tops, well logs, and outcrop data. Based on this geologic model, a simulation grid was developed that includes the Chinle Formation at the base, the Glen Canyon Formation consisting of the Wingate Sandstone, the Kayenta Formation, and our target injection formation the Navajo Sandstone, and the overlying Carmel Formation, a sealing layer. A suite of simulations was performed that modeled storing 50 million tons of anthropogenic CO_2 captured from a single power generation unit at Hunter Power Plant over 27 years. The second suite of simulations modeled the area as a regional GCS site and emissions from all of Utah's major CO2 point-source emitters were injected across the model for 30 years. A systematic reduction of emissions from the power generation sector was simulated by a phased shut down of CO_2 sources from power plants after 30 years of simulation time while maintaining emissions from other heavy industries. Results indicate that this area has a capacity to securely store more than 1.4 billion tons of CO_2, suggesting the complex is an ideal commercial-scale GCS site. This material is based upon work supported by the Department of Energy under Award Number **DE-FE0029280.**

575436 Uncertainty Analysis of Impact of Geochemical Reactions on Forecasting CO_2 Storage at a Depleted Oil Field

Carbon dioxide (CO_2) may be sequestered in the subsurface by many trapping mechanisms, including structural trapping, residual and capillary trapping, solubility trapping, and mineral trapping. While these mechanisms are predicated on geological structures and multiphase fluid flow, geochemical reactions between rock and fluids (CO_2 and original fluid in place, e.g., oil and water) also affect permeability and porosity. The impacts of geochemical reactions on CO_2 storage in deep saline formations is still under intense study by the community; however, its impact on CO_2 storage in depleted oil reservoirs remains unclear.

In this study, we quantify the impact of geochemical reactions during and after CO_2 injection. We selected the Morrow B sandstone at the Farnsworth Unit (FWU) enhanced oil recovery with CO_2 (CO_2-EOR) field in northern Texas for this study. Preliminary results suggest that the impacts of reactive transport on altering properties of porous media are insignificant, at least for a 100-year time frame. By conducting additional uncertainty analysis, we are refining the impacts of geochemical reactions on CO_2 storage forecasting with uncertainty in heterogeneity and geochemical environment taken into consideration.

578206 Impact of Relative Permeability and Capillary Pressure Parameters on Chemical Trapping Predictions in Ccus

Injecting anthropogenic carbon dioxide into the subsurface is considered as a viable option for climate change mitigation. CO_2 injection is also used as a tool in commercial geological operations (e.g. CO_2-enhanced oil recovery, CO_2-enhanced geothermal systems, etc.) for process efficiency enhancement. In any of these applications, it is critical to understand and predict the behavior, migration and long-term fate of the injected CO_2. We investigated the impact of relative permeability and capillary pressure (RP-CP) parameters on the predictions of CO_2 trapping in solution and as minerals. It is difficult to acquire lab-tested RP-CP data specific to the formations being modeled. Moreover, even applying lab-derived RP-CP data in the models can lead to significant uncertainty due to unforeseen heterogeneity in the reservoir. In most cases, generic RP-CP parameters from literature are employed in reactive transport models. Our hypothesis is that this choice of RP-CP parameters can lead to significant differences in the predictions of solubility and mineral trapping made by the models. We investigated four lab-measured RP-CP curves calibrated to the van Genuchten model as applied in most multiphase flow models. Results were compared to those obtained from "standard" parameters used frequently in literature. Significant differences in CO_2 plume migration and dissolution were found in the post-injection phase (>200 years). The reservoir geochemistry was set to that of the Morrow Sandstone of the Farnsworth Unit, Texas. This project was conducted under the auspices of the National Energy Technology Laboratory and its Southwest Regional Partnership on Carbon Sequestration.

575846 Total Internal Reflection Fluorescence Microscopy: Dual Color Imaging on Oil Sands Tailing Ponds

Total Internal Reflection Fluorescence Microscopy: Dual Color Imaging on Oil Sands Tailing Ponds

Abstract:

Understanding the nature of bitumen, clay and their interactions in tailing ponds is essential to support oil industry to develop effective methods for tailing remediation. Clay minerals such as kaolinite has been reported to have negative effect on bitumen extraction and product quality from oil sands. Therefore, understanding of bitumen-clay and clay-clay interaction is crucial in waste water treatment processes such as particle aggregation and sedimentation. In this study, due to natural fluorescent of bitumen, high intensity contrast between clay and bitumen, and high signal-to-noise ratio of Total Internal Reflection Fluorescence (TIRF) microscopy, distribution of both organic bitumen and inorganic clay in water-based bitumen extraction was investigated. Multicolor imaging represented residual hydrophobic bitumen bridging and covering fine clay particles and hydrophilic clay particles in mature fine tailings.

Introduction

The oil sands deposit in the northern Alberta, Canada consists of bitumen (4–18 wt.%), sands (55–80 wt.%), fine solids (particles smaller than <44 μm (5–34 wt.%)) and water (2–15 wt.%) [1]. The semi-solid nature of bitumen requires extraction and recovery technologies different from the conventional processes. These recovery processes result in accumulation of almost one billion cubic metres of a clay slurries known as mature fine tailings (MFT) [2] at 30-35% solid content of fine clays which remain for decades in a fluid state because of its very slow agglomeration rate [2].

Clay component of the oil sands ore is different than many mining industry slurries. Most of the clay fine particles (<2 μm), commonly of aluminum and magnesium, in Athabasca deposits are water-wet with a thin water film of about 2 μm covered the sand grains. Clays are made of layers built from components with tetrahedrally and octahedrally coordinated cations [2]. Kaolinite ($Al_2Si_2O_5(OH)_4$) has an important role because of its high abundance in nature, relatively pure chemical composition, and well characterized crystal [3]. Clays are attached with organic materials at anano-scale in oil sand deposits. Lack of contrast, multistep sample preparation, scale limitation and temperature and pressure dependency limit the study of interaction using conventional methods such as X-ray photoelectron spectroscopy (XPS) [4] and transmission electron microscopy (TEM) [5]. However, due to natural fluorescence of bitumen [6], fluorescence microscopy has shown excellent imaging results over a wide range of spectrum. Fluorescent dye has been extensively used in biomedical diagnosis and biological imaging. A dye with well separated excitation and emission peaks compared with bitumen improves detection of bitumen in MFT.

In this work, spectrofluorometery was first used to verify the favorable bitumen excitation and emission wavelengths for fluorescence microscopy. Moreover, confocal fluorescence microscope was also used to verify bitumen fluorescence in presence of water and clay. According to the results and in agreement with literature, a 488 nm laser source and 500-550 nm emission filter are the optimum values for bitumen fluorescence imaging. A water-soluble fluorescent dye labeled water to make TIRF multicolor images of tailing samples. Successful detection of bitumen in the MFT samples can be achieved with this technique including fluorescence.

Experimental Section

Kaolin, a model clay, and Athabasca bitumen were provided by Institute of Oil Sands Innovation (IOSI), University of Alberta, and MFT was received from Syncrude Canada, Ltd. tailings ponds. MFT samples contain approximately 31 wt. % solids, 4 wt. % bitumen, and 65 wt. % water. Fluorescent dye (CF 660R) was purchased from Biotium Inc. USA, and spectroscopic grade dimethyl sulfoxide was supplied from SigmaAldrich. To observe a real time and condition MFT sample, a bowl-shaped sample holder was prepared using a 10 mm diameter circular hole in the center of a glass slide attached to a cover slip. The sample was placed into the bowl to study using excitation wavelength of 488 nm. The labeling stock solution was diluted in Dimethyl sulfoxide (10 mM) and stored at -20 °C. Water was fluorescence labeled using diluted stock solution (100 times) and mixed with MFT sample.

Spectrofluorometery and TIRF Imaging

Spectrofluorometer was used to measure excitation and emission spectra related to fluorescent dye. A Cary Eclipse fluorescence spectrophotometer was applied to choose the best laser source and emission filter for fluorescent dye observation. A very diluted solution of dye and water was prepared to avoid high excitation light scattering and 4ml of the solution was taken into glass cuvette to measure the excitation and emission spectra.

TIRF microscopy illuminates a thin layer near the interface of two media with different refractive indexes through total internal reflection in the lower index medium at a laser excitation angle greater than critical angle to reflect the generated light. A Nikon eclipse Ti TIRF equipped with 488 nm and 633 nm lasers sources (Melles Griot) were used for excitation and fluorescence in the range 500-550 nm (green) and 650-730 nm (red) were collected. A QuantEM 512sc CCD camera coupled with a 100X objective was used to capture the images.

Result and Discussion

According to figure 1, spectrofluorometer result represented that CF 660R dye has the most excitation and emission in the range of 635-675 nm and 660-700 nm, respectively. Therefore, 633 nm laser line and 650-725 nm filter were used based on availability for the fluorescent microscopy. Bitumen has the most excitation on 488 nm laser line and omits light mostly in 500-550 nm [7]. These fluorescence absorption and emission spectra were set to get the best images of bitumen and dye in MFT samples.

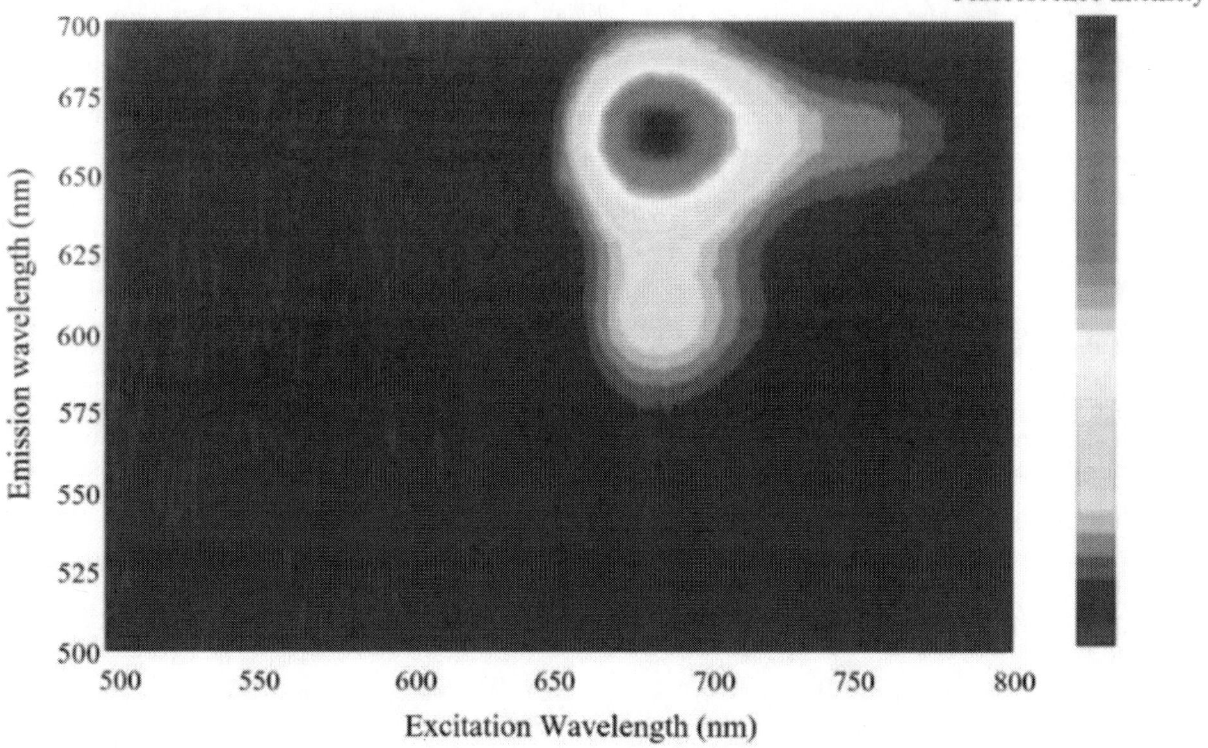

Figure 1. CF 660R spectrofluorometer graph of fluorescence intensity for a range of excitation and emission wavelengths

As discussed earlier, fluorescent dye was used to label water to locate clay, bitumen, and water at the same time in MFT sample. Bitumen, model clay and dyed water multicolor imaging is shown in figure 2. Figure 2a shows a bright-field image of the clay and bitumen interface and the fluorescence image of the clay and bitumen interface with excitation at 488 nm and emission filter in 500-550 nm is represented in figure 2b. Fluorescence image of the clay and bitumen with excitation at 633 nm and emission filter in of 660-700 nm is also shown is figure 2c. Figure 2d is the superimposed 488 nm excitation laser TIRF and 633 nm excitation laser TIRF images using ImageJ® software that shows clay and bitumen interface is clearly visible. The dark black particles in these images are clay particles, while the green dots are bitumen.

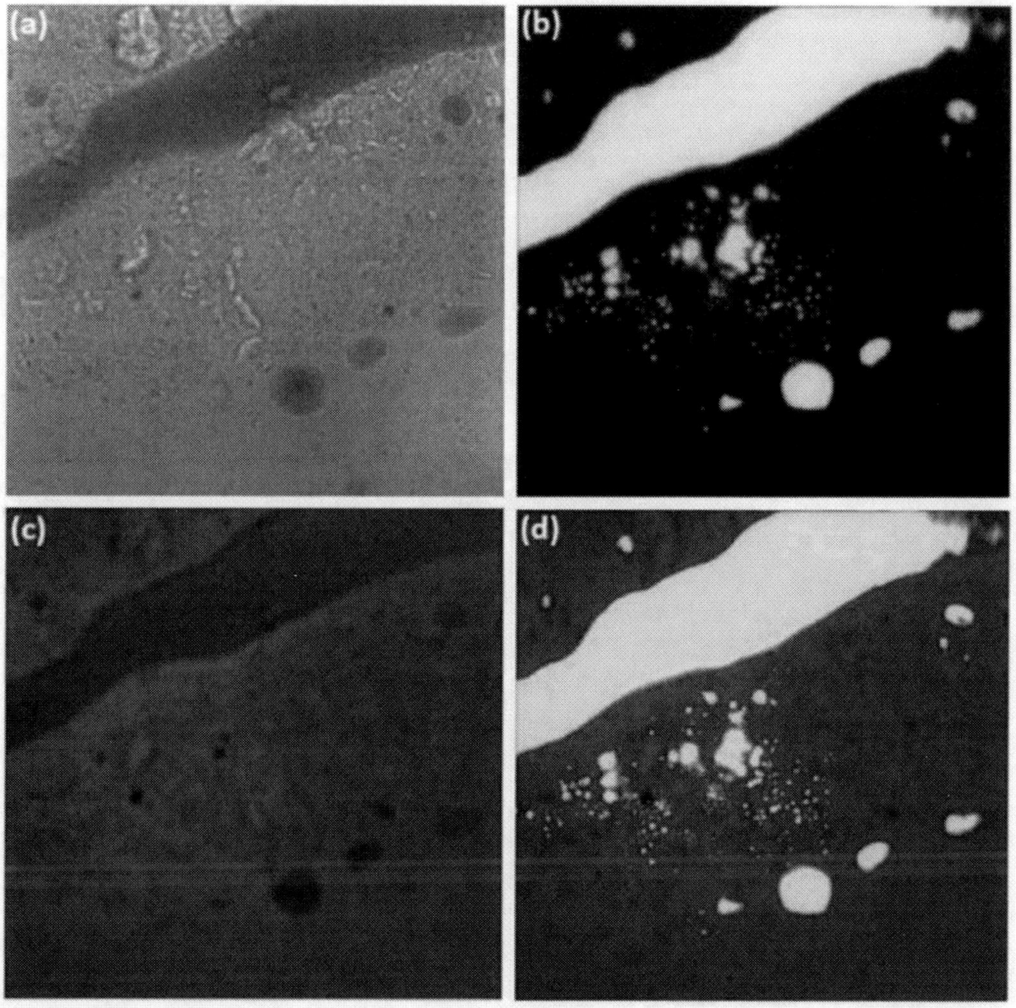

Figure 2. (a) bright-field images of clay-bitumen interface, (b) fluorescence image of the clay-bitumen interface (excitation at 488 nm and emission filter of 500-550 nm), (c) fluorescence image of the clay-bitumen interface (excitation at 633 nm and emission filter of 660-700 nm), (d) superimposed TIRF images of (b) and (c).

A bright-field image of the clay and bitumen interface in a tailing sample is represented in figure 3a. Fluorescence images of the clay and bitumen interface with excitation at 488 nm and emission filter in 500-550 nm and excitation at 633 nm and emission filter in of 660-700 nm are shown in figure 3b and 3c, respectively. Figure 3d is the superimposed 488 nm excitation laser TIRF and 633 nm excitation laser TIRF images to show clear clay and bitumen interface. The dark black particles in these images are clay particles, while the green dots are bitumen.

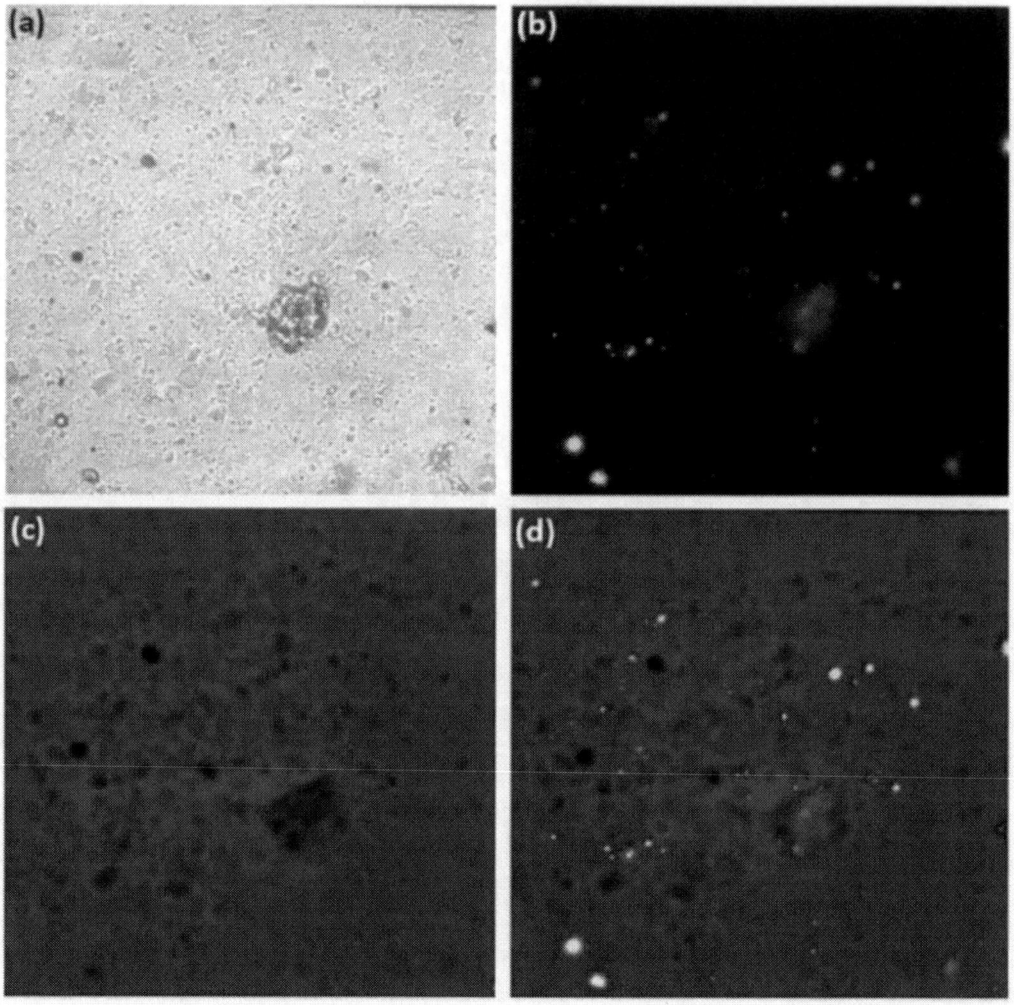

Figure 2. (a) bright-field images of MFT sample, (b) fluorescence image of the MFT sample (excitation at 488 nm and emission filter of 500-550 nm), (c) fluorescence image of the MFT sample (excitation at 633 nm and emission filter of 660-700 nm), (d) superimposed TIRF images of (b) and (c).

Conclusion

The bright-field images of clay and bitumen show that the lack of contrast makes it difficult to observe bitumen. While fluorescence microscopy clearly shows bitumen as a result of its strong fluorescence, clay is still not detectable because it lacks substantial fluorescence. With dye added to this mixture fluorescent images confirm that bitumen and clay can be distinguished using multicolor fluorescence techniques. This is critically important to analyze the bitumen clay interaction to improve effective processes to accelerate aggregation and sedimentation during the MFT treatment. Results indicate that in MFT sample, bitumen is not suspended freely in water and residual bitumen is bridging and coating the clay agglomerates. It also confirms the presence of water-wet clays due to small percentage of clay aggregates attached to bitumen.

References

[1] B.D. Sparks, L.S. Kotlyar, J.B. O'Carroll, K.H. Chung, J. Petrol. Sci. Eng. 39, 417-430, 2003.

[2] N. Romaniuk, M. Mehranfar, H. Selani, B. Ozum, Proceedings Tailings and Mine Waste, Vancouver, Canada, 2015

[3] R. B. Neder, M. Burghammer, TH. Grasl, H. Schulz, A. Bram, S. Fiedler, Clays Clay Mineral. 47, 487–494, 1999.

[4] R. Solc, M. Gerzabek, H. Lischka, Geoderma, 169, 47-54, 2011.

[5] R. Lane, A. Macpherson, S. Magennis, Optics Express, 20, 2012.

[6] F. Handle, J. Fussl, S. Neudl, D. Grosseg, Material Struc. 49, 167-180, 2016.

[7] S. Pendharker, S. Shende, Z. Jacob, N. Nazemifard, Fuel, 207, 262-267, 2017.

574111 Geochemical Characterization of Shale Rocks By Laser Induced Breakdown Spectroscopy (LIBS)

Geochemical characterization of source rocks plays a key role in oil and gas exploration. While elemental composition of shale can provide clues to the rock properties affecting the oil and gas accumulation, mapping of rocks for elemental distribution can be useful to determine mineral phases and chemical zonation within a mineral. In this study, laser induced breakdown spectroscopy (LIBS) was used for the analysis and mapping of shale samples obtained from various locations in the United States. LIBS is a fast analytical technique that can provide elemental information of every element on the Periodic Table (including C, H, N, O) of a given sample. Additionally, LIBS provides depth profiling and can operate in extreme conditions, which makes it an attractive tool for *in-situ* analysis of the bore hole walls at various drilling depths. Shale samples were analyzed using a 266 nm Nd:YAG nanosecond pulse duration laser and a grid pattern that covered 64-144 mm^2 area. The data collected from these experiments were used to construct 2D elemental maps and determine elemental concentration for each sample. The accuracy of analysis was assessed by analyzing the samples of known elemental concentrations. This study shows how the spatial elemental (including C and H) composition varies as a function of the sample location and illustrates the benefits of LIBS over existing analytical techniques.

578041 Experimental Investigation of Synergy between Low Salinity and Polymer Flooding in Enhancing Heavy Oil Recovery on Alaska North Slope

The first ever polymer flood pilot has been initiated to develop the heavy oil (100-5,000 cp) on the Alaska's North Slope (ANS), where thermal methods are not feasible due to various environmental, technical and economic concerns. This study aims to understand the role of salinity during polymer flood. Several striking findings were observed, which are beyond the expectation of the well-established classic polymer flooding philosophy.

Core flooding tests were carried out at reservoir temperature. The crude oil viscosity was 286.3 cp. Residual oil saturation condition was established by flooding the cores with sufficient brine (> 40 pore volumes, salinity = formation brine = 26,700 ppm). High salinity polymer flooding (HSPF, 26,700 ppm) and low salinity polymer flooding (LSPF, 2,500 ppm) were then sequentially carried out. Residual resistance factor was measured after each polymer flooding process. The effluent was monitored to obtain the water cut and oil recovery factor. Microscopy observation was also conducted to examine the effluent. Reproducibility of the results were validated.

Several interesting findings were observed, some of which were unexpected according to traditional polymer flooding theory. 1) The oil recovery was effectively increased (~5 OIIP%) during HSPF despite the oil viscosity was close to 300 cp , which is well beyond the applicable limit according to traditional screening criteria (oil viscosity < 150 cp). It proves that polymer flooding is still valid even beyond the oil viscosity limit. 2) A significant reduction in residual oil saturation by polymer flooding after tens of pore volumes of water flooding. Traditionally, the well-established polymer flooding philosophy believes that the oil recovery is increased by an improved sweep efficiency. There is no change in residual oil saturation (and thus displacement efficiency) compared with water flooding. 3) After sufficient pore volumes of HSPF, further oil saturation reduction and increase in oil recovery could still be achieved by LSPF even with slightly lower viscosity. 4) O/W emulsion was observed at effluent during LSPF, while no emulsion was detected during HSPF. The in-situ emulsification might be one key mechanism to achieve remarkable incremental oil recovery (~13 OIIP%) by LSPF. The results were tested to be reproducible.

This study suggests polymer flood not only increases the sweep efficiency by improving the mobility ratio, it also increases the displacement efficiency. Compared with HSPF, a better performance can be achieved by LSPF, where in-situ emulsification occurs. The study is useful to expand the application limits of polymer flood and find the optimal salinity for the pilot test. More work will be done in the near future to further investigate the EOR mechanisms of LSPF.

577549 NMR Relaxation and Diffusion in Model Organic Shale Using Molecular Simulations

The promising development in shale gas/oil makes it critical to understand fluid properties in pores of shale kerogen. Confinement in nanopores shifts the fluid phase behavior making exploration and production predictions challenging. As the pore shape, pore size and solid composition can all affect the fluid behavior, prediction of fluid properties in shale networks is still under study by many researchers. In this research, nuclear magnetic resonance (NMR) relaxation measurements of fluids in kerogen are analyzed with molecular dynamics (md) simulation, and molecular density functional theory (DFT) to predict hydrocarbon fluid distribution in kerogen.

NMR is extensively used to probe the pore size distribution, fluid distribution, and surface relaxivity of oil and gas reservoirs. However, the interpretation of these NMR-logs still depends on classical Bloembergen-Purcell-Pound (BPP) and Torrey models of NMR relaxation that make important assumptions about the nature of the molecules and its interaction with the environment. These models constrain our ability to interpret NMR logs, thereby hampering development of effective, and environmentally sound, ways to increase hydrocarbon yield. These fundamental limitations also have a bearing in other areas, including in medicine, where MRI is used. We use atomistic molecular dynamics simulations to enhance the understanding and interpretation of NMR relaxation in confined systems of interest in current oil and gas production. In our research team, NMR relaxation behavior of fluids in kerogen from a variety of sources is measured and analyzed with molecular dynamics simulation. Further, DFT is used to predict hydrocarbons absorbed in kerogen matrix and the distribution of hydrocarbons to nano-pores lined with permeable kerogen.

In this talk, we outline how we can use the atomistic simulations to obtain insights into NMR 1H dipole-dipole and spin-rotation relaxation mechanisms. We will discuss our results in modeling the NMR relaxation dynamics and diffusion in bulk hydrocarbons, and then the effect of nano confinement on the same. We conclude by noting some current directions that our team is pursuing.

559703 Determination of Confined Fluid Phase Behavior Using Modified Peng-Robinson Equation of State

Confinement effect of nanopores in shale reservoirs causes the fluids behaving significantly different from bulk fluids. The unique structural and thermodynamic properties of the confined fluids, such as oscillatory density profile and critical property shift, disable the commonly used equations of state (EOS) to be applied in such reservoirs. The objective of this work is to modify the existing Peng-Robinson EOS (PR EOS) to make it applicable for unconventional reservoir fluids.

In this work, techniques have been developed to compute the phase behavior and thermodynamic properties of the confined fluids by coupling critical property shift and capillary pressure in the PR EOS, where a new term representing the molecule-wall interaction resulted from the pore proximity has been introduced. On the basis of data collected from both experiments and molecular simulations, a correlation which is a function of dimensionless pore size has been developed and incorporated in the modified PR EOS to correct the shifted critical properties. The capillary pressure effect is also considered in phase equilibrium computation to account for the strong capillary force in nanopores, where the pressure in vapor phase is larger than that in liquid phase. The newly modified PR EOS has been validated by reproducing the molecular simulation data from a previous literature, yielding an overall error of 7.71%. Both critical property shift and capillary pressure are found to be important impact factors on the phase behavior of confined fluids. The critical property shift is found to cause a significant shrinkage of the phase envelope, whereas the capillary pressure leads to a mild expansion of vapor-liquid two-phase region. The combined effect has been incorporated in the modified PR EOS which has been successfully used to compute the phase behavior of three unconventional reservoir fluids. The bubble-point pressures under reservoir temperatures for Bakken, Eagle Ford, and Wolfcamp fluids at 10 nm are found to be reduced by 31.98%, 23.15%, and 37.77%, respectively.

577099 Optimized Immiscible CO_2 Injections for Enhancing Tight Oil Recovery

Optimized Immiscible CO_2 Injections for Enhancing Tight Oil Recovery

Kaiqiang Zhang, Songyan Li, Na Jia

Keywords: CO_2 enhanced oil recovery; Optimized injection schemes; Tight oil; Immiscible flooding.

In this study, four different injection schemes are specifically optimized through the high-temperature and pressure CO_2 immiscible flooding in tight formations. More specifically, first, several important physical properties of the light crude oil–CO_2 systems are measured, which include the oil density and viscosity as well as the oil–CO_2 IFTs. Then, A total of 10 immiscible coreflood tests are conducted at the injection pressure of P_{inj} = 20.0 MPa and reservoir temperature of T_{res} = 130.0 °C by means of the four different injection schemes, i.e., CO_2 continuous gas injection (CGI), CGI + soaking, pulse injection, and gas-alternating-soaking (GAS). In addition to measured oil recovery factors (ORFs), the pressure difference in each test is monitored and analyzed to achieve a better understanding of the aforementioned four CO_2 injection schemes in the tight formation. Finally, the asphaltene and wax contents in the produced oil are measured at different production periods for each test.

The measured oil viscosity and density are found to be reduced with the temperature increase. Also, the minimum miscibility pressure is measured to be 30.0 MPa at T_{res} = 130.0 °C from the vanishing interfacial tension (VIT) technique. A total of 10 immiscible coreflood tests are conducted to study four different CO_2 injection schemes at the injection pressure of P_{inj} = 20.0 MPa and reservoir temperature of T_{res} = 130.0 °C, i.e., CO_2 CGI, CGI + soaking, pulse injection, and GAS. Moreover, soaking is found to be an important step for the CO_2 enhanced oil recovery at the immiscible conditions. Among all tests with the same experimental conditions, the measured ORF of Test #9, which is applied with the CO_2 GAS at the injection rates of q_{CO2} = 0.5 cm^3/min, is the highest. The CO_2 CGI + soaking at the injection rates of q_{CO2} = 0.5 cm^3/min has the second highest ORF in Test #4. The CO_2 CGI performs better at a small injection rate, whereas the CO_2 pulse injection always gives a low ORF no matter at small or larger injection rate. A small injection rate is suggested to be used for any CO_2 injection scheme. Tests #1, #4, #7, and #9 at the injection rate of q_{CO2} = 0.5 cm^3/min outperform other tests by using the same injection scheme under the same experimental conditions in terms of the ORF but larger injection rates. The smaller injection rate also contributes to delay the CO_2 BT so that a higher ORF can be reached with a lower cost. Last but not least, the asphaltene and wax contents are found to be decreased in the produced oil with the progressing oil production. In the final period of oil production, the asphaltene and wax contents reduce to be 33-51% and 14-25% less in comparison with those at the beginning, respectively.

In comparison with the miscible conditions, most CO_2 enhanced oil recovery (EOR) projects in the shale/tight formations are conducted in the immiscible conditions. However, few studies regarding the evaluations of the CO_2 injection schemes under the immiscible conditions have been conducted through the tight coreflood tests. A series of optimum conditions are determined for different CO_2 injection schemes, which are of paramount importance and can be directly applied for various CO_2 EOR projects in shale/tight formations worldwide.

568924 Produced Water Purification By a Vacuum Flow-through Evaporation

The costs of treating produced water typically outweigh the costs of injection, and as a result 91% of produced water in the United States was disposed of via reinjection. The expansion of oil and gas production and water intensity of hydraulic fracturing methods has increased both the demand for new water supplies and the generation of produced water. The majority of the produced water is disposed of by deep well injection, but recent studies have determined that this induces seismicity. There is a critical need to identify alternative approaches for management of produced waters throughout the United States.

In this work, we demonstrate significantly enhanced water flux and ion rejection rate in ZnO deposited membranes by a vacuum flow-through evaporation. ZnO was deposited on macroporous alumina membranes *via* atomic layer deposition (ALD) to improve water flux by increasing their hydrophilicity and reducing mass transfer resistance through membrane pore channels. The interaction between ZnO ALD layer and water molecules enhanced water flux and ion rejection rates. Compared to uncoated pristine membranes, it was found that ZnO ALD membranes substantially improved water flux while keeping excellent salt rejection rate.

576118 New Nanocatalyst Preparation for Improving Product Quality of in-Situ Upgrading Technology (ISUT)

Exploitation of unconventional energy resources, heavy and extra-heavy oils in particular, is of the essence in order to cope with the ever-growing global energy demand. Yet, complex challenges exist on the pathway of converting these raw hydrocarbons into the form of usable energy and fuels. Due to the low H/C ratio, high concentrations of heteroatoms such as nitrogen, sulphur and metals, the quality of heavy oils is poor and therefore further upgrading processes are required. As a result, both the environmental footprint and production cost of these resources increase appreciably, as compared to the conventional oils. In-reservoir upgrading processes could potentially address these challenges by permanently enhancing the quality of produced oil, which in return redeems the necessity for further processing of the product on the surface. A nanocatalytic in-situ upgrading technology (ISUT) has been recently proposed and showed success in producing premium quality oil, which needs no further processing once reaches the surface [Pereira-Almao et al., Patents CA2864788C and CA2810022C]. In the ISUT process, the heaviest fraction of oil, vacuum residue (VR), is separated aboveground and then is co-injected with hydrogen and nanocatalysts into the reservoir. The hydroprocessing reactions taking place downhole in the reservoir permanently upgrade the oil, which is then recovered on the surface. Then again, the VR is separated and injected back to keep the production cycle continuous. In this process, nickel and molybdenum are used as catalysts and are prepared and dispersed online in the VR. The salts of these transition metals are dissolved in water in order to produce transient microemulsions in the presence of VR, employing an intensive mixing method. Then at high enough temperature, 350 °C, and in the presence of a sulfiding agent, the decomposition of the salts leads to production of catalytically active nanoparticles of metals, i.e. metal sulfides.

A new nanocatalyst preparation pathway using ultrasound is investigated in this study. An ultrasonic probe is employed for preparing the microemulsions of the molybdenum solution with VR. The results of catalyst preparation showed that the synthesized NiMo nanoparticles are half the size of the particles prepared conventionally using a high shear mixer. A continuous setup which potentially emulates a patented process (US20180086990A1) for catalyst preparation for the ISUT application to an actual sandstone reservoir was constructed to evaluate the two different batches of nanoparticle suspensions in VR, one using high shear mixers, and one using ultrasonic probe. A reactor was packed with silica-based sand with measured permeability of 15-16 Darcy, which effectively resembles Canadian oilsand reservoirs. In the continuous setup, a comparative study was done to assess the reactivity of the two prepared batches. The VR along with hydrogen and catalyst were injected in the reactor and the residence time was 48 hr for all experiments.

Significant improvement in the product quality was observed once the ultrasound-assisted nanoparticles were used. Using the new preparation approach, lower density and viscosity, lower microcarbon content and greater demetallization and desulphurization extent were achieved for the products. Therefore, it was concluded that employing the ultrasonic pathway for catalyst preparation in ISUT processing improves the overall upgrading efficiency of the process. As well, smaller size nanoparticles ensure that the permeability of the sandstone remains intact since there is less chance of pore impairment in the case of smaller particles. Also, after the experiment, the sandpack was carefully cut in pieces and the sand in each part was characterized in terms of the catalyst concentration and the nature of the remaining hydrocarbon. It was observed that almost all of the catalytic precursors were deposited in the first one-fourth of the reactor. In addition, this first section showed to have heavier hydrocarbons remaining in the reactor.

566326 Regional Economic Impacts of Carbon Capture and Sequestration

Regional economic impacts are a critical piece of the decision process surrounding carbon capture, utilization, and sequestration (CCUS) implementation. Integration of carbon capture technologies with the existing coal fleet not only helps to ensure the longevity of an existing industry, jobs are created at the capture plant during the construction and operational periods. A significant number of jobs are created during the construction period, which results in additional job creation through indirect and induced effects. Local suppliers will see an uptick in sales due to the high demand for construction materials, which is considered an indirect impact on the local economy. Workers spend money on housing, food, and entertainment that can induce additional job creation. The ripple effects of a project such as this can last for many years and result in increased tax revenue to the state and other localities. Permanent jobs are created during the operational phase at both the plant and oil wells that are ready for enhanced oil recovery (EOR) operations. The incremental oil production can also increase tax revenues to the state. This presentation will review the potential regional economic impact of wide-scale CCUS technology implementation on coal-fired power plants and oil fields in North Dakota.

The economic impacts of activity in the energy sector can be determined through the use of input–output (I-O) models. There are many commercially available I-O software models; the three most commonly used software models for economic impact modeling are the Regional Input–Output Modeling System (RIMS-II), Impact Analysis for Planning (IMPLAN) and Regional Economic Models, Inc. (REMI). All of the I-O analysis models produce the economic impacts of three basic effects—direct effect, indirect effect, and induced effect—which, when totaled, are equivalent to the total economic impact/change based on the circumstance modeled. The direct effect is essentially the amount of spending as a direct result of the change in the industry sector under study. For the purposes of this study, this would be the amount of capital expenditure for the construction of a CO_2 capture system at a power plant and the construction of the pipeline required to transport the CO_2 to the oil fields for EOR activity. This direct effect would include not only the capital plant expenditures but the salaries paid to the employees by the companies building the plant. The indirect effect would be the amount of economic impact on the state from the contractors and subcontractors purchasing goods from local/state entities such as steel, concrete, welding supplies, etc. The induced effect is the change in economic impacts due to employee spending of earned income for restaurants, clothing, cars, visits to doctors, etc.

The EERC has evaluated the economic impact of CCUS technology implementation in the state of North Dakota. The estimated capital and operational costs for these activities have been determined and then analyzed using the IMPLAN model with region-specific modifications. The direct, indirect, and induced effects have been determined for a number of scenarios. This presentation will review the results of these studies and show the positive regional economic impact that can be realized when CCUS technologies are deployed.

553642 Using CO_2-Responsive Nanogel to Improve CO_2 Flooding Sweep Efficiency and CO_2 Storage Efficiency

Using CO_2-Responsive Nano-gel to Improve CO_2 Flooding Sweep Efficiency and CO_2 Storage Efficiency

Introduction

Due to the size limitation, millimeter-sized particle gels can only penetrate in the fractures or channels, which are mainly focused on the near wellbore problems. However, the in-depth plugging cannot be achieved by conventional millimeter-sized particle gel treatment. Therefore, the nano-particle gels are developed based on the demand of the in-depth plugging. In this study, a novel nano-gel with CO2 responsive property is proposed for the first time. The CO2-responsive nano-gel can be stimulated by CO2 to increase particle size; therefore to provide additional plugging to CO2 flooding sweep efficiency and CO2 storage efficiency.

The novel CO_2-responsive nano-particle gel was examined using sandstone cores in order to study the transportation behavior and whether it can form an in-depth plugging during CO_2 flooding. Six cores were used in this experiment and were divided into two groups: one is the group with CO_2 and another is the group without CO_2. The permeability of the core is 40, 130, and 370 md, respectively. The plugging efficiency of the group without CO_2 was measured directly after nano-gel injection, while the plugging efficiency of the other group was measured after CO_2 stimulation in the high-pressure vessels using brine. Transportation test of nano-gel was conducted using homogeneous sandstone cores. Five pressure sensors are installed along the core evenly in order to monitor the nano-gel transportation. The nano-gel was injected into the core until the third pressure sensor has reading, then the brine is injected into the core until the fourth pressure sensor has reading. After placing the nano-gel in the middle of the core, supercritical CO_2 was injected into the core until CO_2 breakthrough from the outlet. The core was then soaked in supercritical CO_2 for 7 days to allow the interaction between the nano-gel and supercritical CO_2. The plugging behavior was tested by the following water flooding

1. Materials

1.1 CO_2-responsive nano-gel

CO_2-responsive nano-gel has a novel function that it can increase its swelling ratio under CO_2 condition. Therefore, after placing the nano-gel in the in-depth, the nano-gel can swell more during the post CO_2 flooding, hereby divert the CO_2 to unswept area to increase oil recovery. In this experiment, the concentration of CO_2-responsive nano-gel is 2,000 ppm and the nano-gel solution is prepared with 1% NaCl brine.

1.2 Sandstone rock

The sandstone rock used in this study has a permeability around 100 md. The diameter of the core is 1.5 inch and the length is 1 ft.

1.3 Supercritical CO_2

The CO_2 used in this experiment is in supercritical CO_2 condition. The CO_2 is pressurized in an accumulator which is wrapped with a heating pad, therefore, the pressure and the temperature is higher than the CO_2 critical point to achieve supercritical condition.

2. Experimental apparatus design

The Figure 1 is the experimental apparatus design. The sandstone core is put into a coreholder and secured by adding confining pressure. Three pressure sensors are installed along the coreholder evenly and two pressure sensors at inlet and outlet, respectively. The pressure data is transmitted to the computer. The whole system is heat to 45¡æ. The back pressure regulator (BPR) is used to increase the whole system pressure above CO_2 supercritical point. Keeping both temperature and pressure above the supercritical point can ensure that the CO_2 flow through the system is in the supercritical condition. The nitrogen gas source is used to provide back pressure. The nano-gel and supercritical CO_2 are injected using two accumulators.

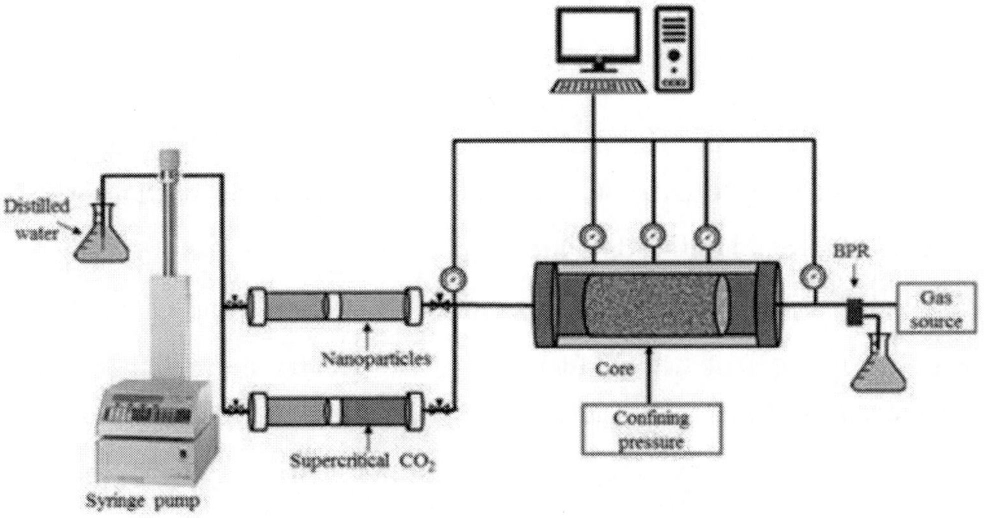

Figure 1 Experimental apparatus design

3. Experimental procedures

3.1 Nano-gel preparation

The nano-gel concentration is 2,000 ppm and the solvent is 1% NaCl brine. After dissolve the nano-gel in the brine, the solution is put into 65¡æ oven for 24 hours to allow thenano-gel reach an equilibrium swelling ratio. The nano-gel is filtered using 10 μm filter paper before use to remove any impurities.

3.2 Core flooding procedures

1) Measure the dry weight of the core. Vacuum the core and saturate the core with 1% NaCl brine;

2) Measure core wet weight to determine the pore volume (PV);

3) Set back pressure at 1100 psi;

4) Measure sandstone core permeability with 1% NaCl brine using flowrates of 1, 2, 3, 4 cc/min;

5) Inject 0.5 PV of 2000 ppm nano-gel at 0.05 cc/min (1ft/day) flowrate (or inject nano-gel until the second pressure sensor has reading);

6) Inject 0.25 PV of 1% NaCl brine at 0.05 cc/min flowrate (or inject brine until the third pressure sensor has a reading);

7) Inject supercritical CO_2 using 4 cc/min at 65 ¡æ until CO_2 breakthrough from the outlet;

8) Keep the CO_2 accumulator connected with the coreholder and soak the core in supercritical CO_2 for 7 days;

9) Inject water using 1, 2, 3, 4 cc/min flowrates to test the plugging efficiency.

4. Results

4.1 CO_2-responsive nano-gel plugging efficiency

As shown in Figure 2, nano-particle gel has the best plugging performance in low permeability cores after CO_2 stimulation. After CO_2 stimulation, the residual resistant factors (Frr) of most of the cores were improved, which represents that the novel CO_2-responsive nano-gel plugging efficiency can be improved in supercritical CO_2 condition. The plugging efficiency of nano-particle gel to brine is plotted in Figure 4.5. The results show that the maximum plugging efficiency is achieved in 40md and it decreases with the increase of rock permeability. After CO_2 stimulation, the plugging efficiency to brine is more than 90% in the 40 md core.

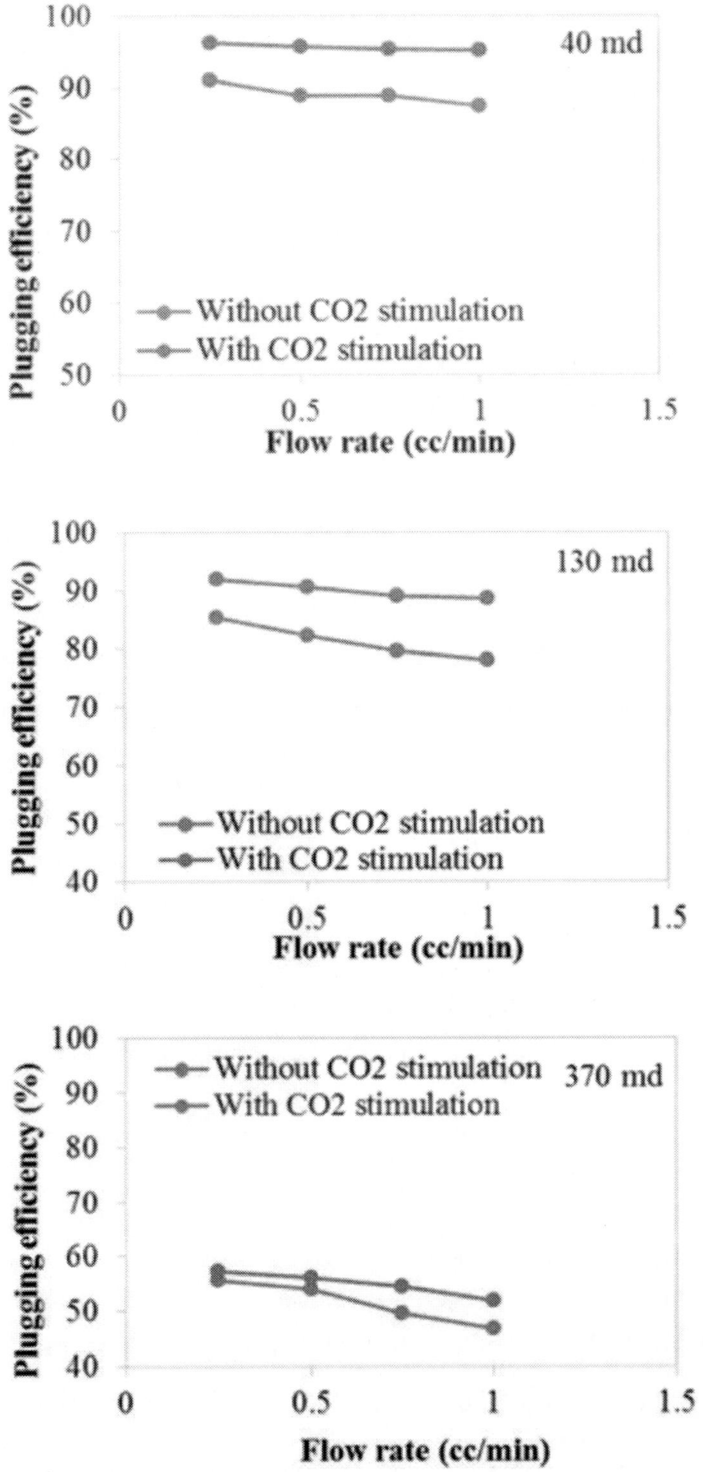

Figure 2 Plugging efficiency of nano-particle gels to brine

4.2 CO_2-responsive nano-gel transportation in sandstone rock

4.2.1 Absolute permeability measurement and homogeneity test

As shown in Figure 3, the differential pressure for each segment is almost the same, which indicates that this sandstone core is nearly homogenous. The water absolute permeability for this sandstone core is 98 md.

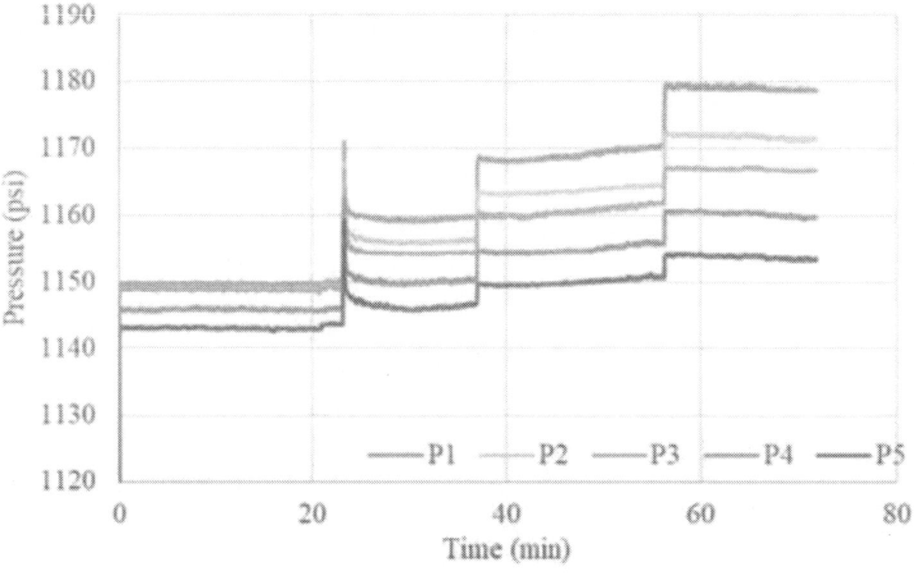

Figure 3 Absolute permeability measurement results

4.2.2 Nano-gel injection process and post water flooding process

0.43 pore volume (PV) of nano-gel was injected at a constant flowrate of 0.05 cc/min. As shown in the figure below, the differential pressure for each segment is almost the same throughout the nano-gel injection process, which indicates that the 0.43 PV of nano-gel injection did not result in an efficient plugging before CO_2 stimulation. Therefore, the nano-gel can transport into the in-depth without forming a surface plug or near wellbore plug.

After placing the nano-gel in the core, a 0.25 PV brine was injected into the core in order to displace the nano-gel near the inlet and push the nano-gel moving towards the in-depth.

4.2.3 CO_2 injection and soaking process

The supercritical CO_2 was pressurized and heated in an accumulator. The CO_2 was injected into the core at a constant flowrate of 4 cc/min until the CO_2 breakthrough the core. The injection was terminated once the CO_2 production was observed. Since the supercritical CO_2 broken through the core, the contact area between the CO_2 and the nano-gel increased, therefore, the nano-gel stimulation under CO_2 condition was more efficient. The core was then soaked in the supercritical CO_2 by keep connecting the coreholder with the CO_2 accumulator for 7 days. The post waterflooding will be conducted to test the plugging efficiency of the nano-gel after CO_2 stimulation.

4.2.4 Post waterflooding process

After soaking the core for 7 days, 1% NaCl brine was injected into the core at various flowrates. The results indicate that the brine injection pressure increased after nano-gel treatment, which revealed that the nano-gel can provide a sufficient plugging to brine after soaking in supercritical CO_2. The residual resistant factor for each segment and the whole core was calculated and plotted in Figure 4. As shown in the figure, the segment 2 has the highest resistance to water flow, while the Frr for segment 3 and 4 was close to 1, which indicated that the nano-gel was not penetrate into segment 3 and 4 or did not form a sufficient plugging.

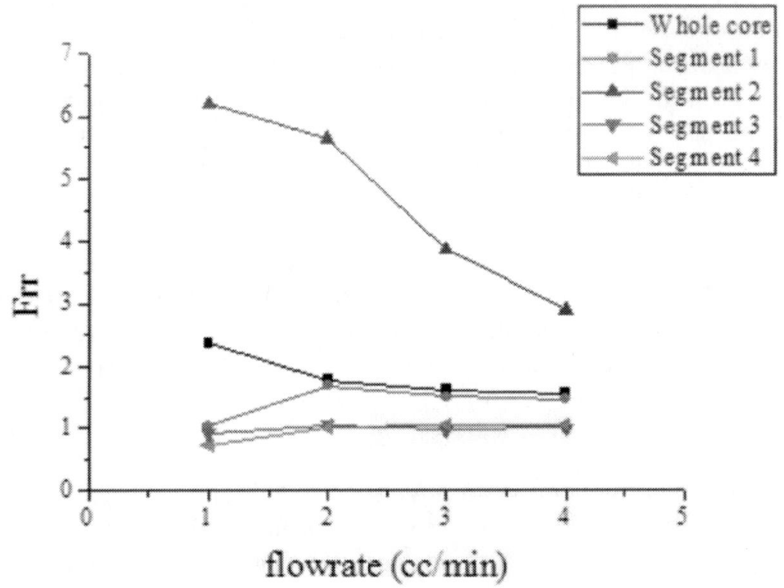

Figure 4 Water residual resistant factor for each segment after nano-gel treatment

Acknowledgement

The authors would like to express their gratitude for the financial support provided by the DOE under contract DE-FE0024558.

575126 Improved Design of a Diesel-Driven Solid Oxide Fuel Cell System for Auxiliary Power Unit Applications

In this study, a diesel-driven solid oxide fuel cell system has been developed for auxiliary power unit applications. Among various hydrocarbon fuels, diesel fuel has many advantages such as high energy density and versatility, and it is often used for heavy-duty applications and remote areas. Diesel itself can be used to generate electricity with a diesel generator, but it produces high pollution and noise, with low energy efficiency. In another way, a diesel autothermal reformer can be introduced to generate hydrogen-rich reformate that can be used for a solid oxide fuel cell stack to generate electricity. With the diesel reformer, the diesel fuel inside a fuel tank of the heavy-duty truck can be directly used to generate hydrogen for the fuel cell system.

With the integration of the diesel autothermal reformer and the solid oxide fuel cell stack, a fuel cell system for the auxiliary power unit application has been developed. For the heavy-duty truck applications, the power output target of the auxiliary power unit was set to one kilowatt to satisfy the electrical needs. Structured catalysts were introduced for the diesel reformer to ensure the mechanical strength of the component. During the development of the diesel reformer, a long-term test for the catalyst component was conducted to ensure its catalytic stability. 2,000 hours of operation was achieved with the structured catalyst for the autothermal reforming.

In commercial diesel fuel, a small amount of sulfur content exists. To protect the fuel cell stack, desulfurization with ZnO based adsorbent was followed by the autothermal reforming process. The temperature of the reformate has to be controlled within the range of the operating window of the desulfurizer to reduce the sulfur content to ppm level. A heat network analysis for the reformer module including the desulfurization process was done and required specifications for heat exchangers were obtained.

Two versions of the integrated fuel cell system were designed with different assumptions for the thermal stability of the solid oxide fuel cell stack. The solid oxide fuel cell stack needs a high operating temperature around 700 deg C, and the thermal stability of the stack determines overall system start-up time. The first version of the system was designed with meeting the requirement of the slow heat-up of the stack. The temperature of the fuel cell stack is increased with the heat from the reformate oxidation in the first design, so it takes some time to reach the operating temperature. However, the start-up time of the fuel cell system can be shortened if the fuel cell stack has enough thermal stability that can sustain the fast heat-up process. Therefore, the system configuration can also be modified at the improved version by using external heat sources during the start-up process. In result, shorter start-up time and simplified start-up process were obtained with the enhanced design of the fuel cell system.

565496 Hydrogen Production from Carbonaceous Feedstocks Using a Novel Solar Rotary Kiln Reactor

HYDROGEN PRODUCTION FROM CARBONACEOUS FEEDSTOCKS USING A NOVEL SOLAR ROTARY KILN REACTOR

Summary

Synthesis gas production using solar energy as a means to valorize carbonaceous feedstocks is an interesting source of hydrogen and an effective alternative to hybridize existing thermal power plants operating with fossil fuels. Some of the benefits of providing solar heat to run endothermic processes such as pyrolysis and gasification are: (i) no combustion products in the resulting gas; (ii) storage of solar energy as chemical energy; and (iii) energy upgrading of the feedstock, among others.

The present work presents the design of a novel solar rotary kiln tilted and bottom indirectly irradiated for the thermochemical conversion of carbonaceous feedstocks, as well as the hydrogen yield obtained from the gasification of bituminous coal. A peak hydrogen concentration of 26% was found in the product gases and a maximum temperature of 1,200 K was achieved in the reaction zone.

---

1. Introduction

Solar synthesis gas (syngas) production from carbonaceous feedstock is a desirable technology to develop in Chile, since it would represent an alternative on foreign imports of fossil fuels and would provide a path to valorize, initially, national sub-bituminous coal reserves, and other carbonaceous feedstocks, such as agroindustrial and municipal solid wastes.

Previous studies by Piatkowski et al have demonstrated the feasibility of converting several feedstocks into syngas using solar energy, achieving energy upgrades by factors up to 1.30. However, the high temperatures required for gasifying solid fuels, as well as the poor thermal conductivity of packed beds Wieckert et al, make the development of a continuous reactor an engineering challenge.

This study was focused on the design, development and operation of a novel solar rotary reactor for thermochemical conversion of coal. The conversion processes observed consisted of pyrolysis, allothermal steam gasification and autothermal steam gasification.

2. Methods and experimental apparatus

Experiments on solar thermochemical conversion were conducted in a batch rotary kiln solar reactor. Rotary kilns for solar applications have been widely reviewed by Alonso et al, but no mention of the design proposed in this research have been made previously. The concentration optics used for the experiments consisted of a 10 heliostats array (1 m^2 of reflective surface each one) focused to a 1.5 m^2 Cassegrain concentrator, with its hyperbolic focal point located behind its parabolic vertex. The reactor featured a cavity receiver, with its entry located on the hyperbolic focal point of the concentrator, coupled to a CPC-3D, responsible of redirecting the radiation spillage into the cavity receiver. Inside the cavity an emitter plate of graphite coated with SiC was used to reradiate the solar flux to the reaction chamber. Steam was supplied to the reaction chamber using a custom-built electric boiler. Coal particles showed a uniform size distribution with an average particle size of 5 mm. Alongside with the fuel particles, inert alumina hollow spheres of similar size were mixed in order to enhance the effective thermal conductivity of the bed. The product gas was cooled and filtered through a stage filled with water and collected vaccum sealed vessel in order to control the total amount of product gas obtained from each gasified sample.

The reactor was implemented at the Renewable Energy Laboratory of the Department of Mechanical Engineering from the UTFSM, and all experiments featured irradiances greater than 1,000 W m^{-2}. Synthesis gas was determined by discrete sampling of the product gas and gas chromatograph analysis. Temperature measurements were achieved using type-S thermocouples uniformely distributed along the reactor walls.

3. Results

Solar thermochemical conversion of bituminous coal was achieved under the presence of steam at a fixed tilt angle of 45°. Peak hydrogen concentration found in the product gases was of 27%. Peak temperature recorded inside the reaction chamber was 1,200 K.

4. Discussion

Solar assisted hydrogen production, was successfully achieved in a novel bottom indirectly irradiated solar rotary kiln reactor operating with bituminous coal and steam. The emitter plate, as recommended by Piatkowski et al, effectively transferred the solar heat to the reaction chamber without signs of damage. Finally, a design for a continuous reactor should consider an ash collection system.

5. References

Alonso, E., Gallo, A., Roldán, M. I., Pérez-Rábago, C. A. & Fuentealba, E. Use of rotary kilns for solar thermal applications: Review of developed studies and analysis of their potential. *Sol. Energy* **144,** 90–104 (2017).

Piatkowski, N., Wieckert, C., Weimer, A. W., Steinfeld, A. Solar-driven gasification of carbonaceous feedstock—a review. *Energy Environ. Sci.* **4,** 73–82 (2011).

Wieckert, C., Obrist, A., Zedtwitz, P. Von, Maag, G. & Steinfeld, A. Syngas production by thermochemical gasification of carbonaceous waste materials in a 150 kWth Packed-Bed Solar Reactor. *Energy and Fuels* **27,** 4770–4776 (2013).

557667 Development of a Dual Fluidized Bed Reactor for Solar Thermal Storage

A dual fluidized bed system can easily integrate a concentrated solar thermal(CST) process and thermo-chemical conversion processes because fluidized bed materials can be used as heat carriers and storage matters. However, the design and the operation of dual fluidized bed reactors is very difficult because the dual fluidized bed reactor is generally composed of a circulating fluidized bed reactor and a bubbling fluidized bed reactor. Therefore, the hydrodynamics characteristics of bed materials in the dual fluidized bed system should be investigated and analyzed to design or develop a solar hybridized dual fluidized bed system. In this study, the attrition characteristics and fluidization characteristics of solar thermal matters used for the dual fluidized bed system is investigated and analyzed by using the ASTM D 5757 attrition tester and a computational fluid dynamics (CFD) model built with Barracuda Virtual Reactor® using the MP-PIC method to simulate the hydrodynamics of a solar hybridized dual fluidized bed system. The CFD model for a solar hybridized dual fluidized bed system consisted of riser, cyclone, hopper, bubbling fluidized bed CST reactor, and loop-seal. It was assumed that air was injected to riser, bubbling fluidized CST reactor, and loop-seal via perforated plates.

573005 Development of a High-Purity Hydrogen Production Skid Unit for Hydrogen Refueling Stations

A high-purity hydrogen (99.999 %) production skid unit (200 kg hydrogen/day, called as *KIER HyPU-100*), comprising natural gas-steam reforming, water-gas shift reactors and vacuum pressure-swing adsorption (VPSA) process, for hydrogen refueling stations has been developed. The system should satisfy the main prerequisites for actual application such as 1) higher than 80 % hydrogen production efficiency on the basis of higher heating values (HHV) and 2) lower than 1 ppm of CO concentration at the final outlet of the hydrogen refining process. In order to guarantee the high efficiency, the heat exchanging network in the reforming process was optimized by the commercial process simulation software package and the heat and mass transport phenomena in the reformer were also intensively studied by the computational fluid dynamic models. In addition, the optimal VPSA process has also been developed to purify the hydrogen from the product gas of the water-gas shift reactor. The performance of the VPSA is not only related to the hydrogen purity but also to the hydrogen production efficiency since the recovery ratio of high-purity hydrogen to the total reformate affects the amount of both product hydrogen and make-up natural gas fed into the burner of the reformer. The developed system showed consistently reliable performance within the long-term test sessions (hydrogen production efficiency: > 80 $\%_{HHV}$, hydrogen purity: 99.999 %) and the operating data was compared with the mathematically predicted values to prove the feasibility of the design concepts.

579746 Hydrogen Isotope Separation By a Palladium-Based Membrane

The separation of hydrogen isotopes is an important technique for many technologies and scientific applications. Hydrogen isotope separation may be achieved using a membrane because of permeability differences between protium (H) and deuterium (D). The membrane permeability is proportional to the product of the solubility and diffusivity of the hydrogen/palladium system. The permeability of H and D were evaluated from a gas mixture containing 4 vol % D2 – 96 vol% H at temperatures 293 K – 473 K in a continuous membrane reactor. Based on the solubilities and diffusivities of the hydrogen isotopes, protium will move through the membrane faster than deuterium. This difference in hydrogen isotope permeability results in enrichment of deuterium on the retentate side of the membrane and protium on the permeate side of the membrane.

569246 Adsorptive Desulfurization of Liquid Fuels at Elevated Temperatures Using Metal Exchanged Zeolite Y

Introduction

The high energy density, ease of transportation, and ready availability of liquid hydrocarbon fuels make them an attractive option for generating electricity remotely with auxiliary power units (APUs) containing a reformer and fuel cells [1]. This approach remains problematic, however, since the sulfur originally in the liquid fuels can quickly deactivate the fuel cell electrodes [2]. As a consequence, ultra-low sulfur fuels (ideally with total sulfur content of less than 1 ppm) are required for the trouble-free operation of fuel cells. Among the technologies available for desulfurizing fuels, adsorptive desulfurization has emerged as a promising approach because it does not require high volumes of solvents or reactive gases, which are not available in remote locations [3]. Therefore, this process has been under investigation for remote desulfurization of liquid fuels for fuel cell applications. More specifically, desulfurization of JP-8 has gained attention because JP-8 is the main fuel used by the military of NATO nations. Generating electricity with fuel cells in remote areas is a primary need for the military, particularly for silent watch and ground soldier applications [1]. JP-8 is composed primarily of paraffinic, aromatic and naphthenic hydrocarbons. Sulfur concentrations in JP-8 can reach up to 3,000ppmw, present predominantly in alkylated benzothiophenes [4].

Several studies have reported using metal exchanged zeolites for adsorptive desulfurization at near ambient temperatures [5], [6]. Since sulfur removal at low temperature involves reversible complexation, spent adsorbents can be easily regenerated [7]. Such weak reversible complexation bonds, however, are not sufficiently specific to remove substantial amounts of sulfur from real fuels because of the co-extraction of aromatic compounds. Aromatics are similar in size and structure to the predominant sulfur species found in jet fuels (alkylated benzothiophenes) and compete with them for the active sites [8]. And since jet fuels usually contain 10%vol to 25%vol of aromatics [9], only a small fraction of the organosulfur compounds present in the fuel can be removed.

Therefore, alternative approaches should be considered to address the problem of competitive adsorption and improve the selectivity of sulfur adsorption. Some theoretical studies have suggested that elevated temperatures may improve the adsorption selectivity of organosulfur compounds on zeolites [10]. However, very little is known about the effect of temperature on the adsorption of organosulfur compounds from liquid fuels on zeolitic materials beyond near ambient temperatures. Our understanding about the effect of temperature on adsorptive desulfurization is incomplete.

Elevated Temperatures Significantly Improve the Sulfur-Removal Capacity of CuNa-Y Zeolite

In an effort to overcome the limitations imposed by the co-extraction of aromatics in liquid fuels, we studied adsorptive desulfurization at temperatures substantially higher than ambient temperature. Na-Y and Cu-exchanged Na-Y (CuNa-Y) zeolites were used to remove sulfur first from model fuels and then from a JP-8 fuel at temperatures up to 180°C.

Batch desulfurization and TPD experiments with model fuels containing 3-methyl-benzothiophene (3-MBT), dodecane and toluene showed that 3-MBT removal was strongly dependent on treatment temperature and involved weak physisorption bonds with Na sites, weak complexation bond with Cu sites at 30 or 80°C and strong chemisorption S-Cu bonds at 130 or 180°C. Overcoming the competition that favors the adsorption of aromatics at low temperatures, the formation of S-Cu bonds at high temperatures shifted the balance and allowed significant levels of 3-MBT removal even when toluene was present at high concentrations.

We then proceeded to carry out desulfurization experiments with a JP-8 fuel containing 2,230 ppmw of total sulfur. While our CuNa-Y zeolite removed very little sulfur from the JP-8 at 30 or 80°C, the use of elevated temperatures dramatically improved the desulfurization efficacy of CuNa-Y zeolite, increasing its sulfur-removal capacity from 2.5mg-S/g-adsorbent at 30°C to 36mg-S/g-adsorbent at 180°C.

To determine if CuNa-Y zeolite at 180°C can remove all sulfur compounds present in JP-8, we submitted the fuel to a sequence of batch desulfurization experiments at this temperature. Treated JP-8 fuel (recovered from the first desulfurization step) was desulfurized again with fresh adsorbent in the batch reactor and the same process was repeated two more times . These sequential batch experiments simulate a continuous desulfurization process. We found that the first desulfurization step removed the least refractory organosulfur compounds, that is those with the lowest boiling points. Subsequent steps progressively adsorbed the more recalcitrant organosulfur compounds until all the sulfur had been removed from the JP-8 fuel.

Our results demonstrate the feasibility of using zeolitic adsorbents and elevated temperatures for ultra-deep desulfurization of JP-8 fuels. This technology could potentially enable the use of liquid fuels such as JP-8 on remotely located integrated auxiliary power units containing a reformer and fuel cells.

REFERENCES

[1] D. T. Tran, "Desulfurization of JP-8 jet fuel : challenges and adsorptive materials," *RSC Adv.*, pp. 7301–7314, 2018.

[2] K. Liu, C. Song, and S. Velu, *Hydrogen and Syngas Production and Purification Technologies.* Hoboken, NJ: John Wiley & Sons, Inc., 2009.

[3] A. Stanislaus, A. Marafi, and M. S. Rana, "Recent advances in the science and technology of ultra low sulfur diesel (ULSD) production," *Catal. Today*, vol. 153, no. 1–2, pp. 1–68, 2010.

[4] D. D. Link and P. Zandhuis, "The distribution of sulfur compounds in hydrotreated jet fuels : Implications for obtaining low-sulfur petroleum fractions," vol. 85, pp. 451–455, 2006.

[5] S. Velu, C. Song, M. Namazian, S. Sethuraman, and G. Venkataraman, "Desulfurization of JP-8 Jet Fuel by Selective Adsorption over a Ni-based Adsorbent for Micro Solid Oxide Fuel Cells," pp. 1116–1125, 2005.

[6] A. J. Hernández-Maldonado, F. H. Yang, G. Qi, and R. T. Yang, "Desulfurization of transportation fuels by pi-complexation sorbents: Cu(I)-, Ni(II)-, and Zn(II)-zeolites," *Appl. Catal. B Environ.*, vol. 56, no. 1–2 SPEC. ISS., pp. 111–126, 2005.

[7] A. J. Herna and R. T. Yang, "Desulfurization of Liquid Fuels by Adsorption via π Complexation with Cu(I)-Y and Ag-Y Zeolites," no. I, pp. 123–129, 2003.

[8] L. Wang, B. Sun, F. H. Yang, and R. T. Yang, "Effects of aromatics on desulfurization of liquid fuel by p -complexation and carbon adsorbents," *Chem. Eng. Sci.*, vol. 73, pp. 208–217, 2012.

[9] H. T. Mayfield, "JP-8 Composition and variability," EQL, Armstrong Laboratory. Approved for public release. 1996.

[10] S. Dang *et al.*, "Competitive adsorption mechanism of thiophene with benzene in FAU zeolite: The role of displacement," *Chem. Eng. J.*, vol. 328, pp. 172–185, 2017.

579794 Numerical Modeling on Water and Heat Recovery from Combustion Flue Gas

A numerical modeling has been developed to simulate heat and mass transfer in condensing heat exchangers in conjunction with R-134a as coolant. The objective of the research is to develop an advanced organic rankine cycle systems to recover both heat and water from boiler flue gas for fossil-fuel power plant applications. One of critical paths to accomplish the research goal is to model simultaneous condensation of water vapors in flue gas side and flow boiling of R-134a in coolant side in a heat exchanger. An approach combining iterative solution technique and finite difference method has been applied to the numerical scheme in the recovery modeling by assuming phase of refrigerant at coolant exit under given or simulated boundary conditions. The simulated boundary conditions have been built up by using the heat and mass balance software, IPSEpro. In this presentation, the simulation results on a pilot-scale organic rankine cycle and phase transitions will be introduced.

569932 Optimal Selection and Design of Energy Storage Technologies Integrated with Advanced Energy Plants

With the increasing interest and focus on renewable energy sources, various fossil-fueled (e.g., coal-, natural gas- and oil-fired) power plants that were designed to operate at base-load conditions are being forced to cycle their load. Cycling includes load-following (ramping of the power generated up/down), shutdown and startup, and operation under variable and minimum load conditions. Increased renewable penetration creates a need for power plants to incorporate fast cycling, fast ramp rates and flexible operation to ensure grid reliability. Cycling leads to high thermal stresses and high failure rates in critical components [1]. To this end, suitable integration of energy storage facilities with the power plants can be instrumental in achieving near-steady state operation of these power plants. Most importantly, decentralized deployment of energy storage at the power plant level can exploit the existing equipment items and facilities at the host power plant for improving the efficiency of the storage technologies and reducing the storage capacity. However, realization of these benefits will critically depend on novel configuration/integration strategies with the least impact on the power plant operation and its configuration. Most of the previous work in integrating energy storage in power generation considers large-scale storage at the grid level, resulting in high storage capacities and costs [2-4]. Furthermore, the storage technologies differ widely based on the maturity of technologies, their costs, size, life, availability, efficiency, footprint, safety and environmental hazards. The grid scale imbalance can significantly vary based on the location, time of the day and the year. Lastly, dynamics of the entire integrated system including both the power plant and the storage technologies must be taken into account to obtain the cost-optimal solution.

In this work, we develop a mathematical optimization-based methodology for downselecting technological choices for decentralized energy storage with the existing power plants. Dynamic models of several electro-chemical, mechanical, and thermal storage technologies are developed along with the dynamics models of the host fossil-fueled plants. A mixed integer nonlinear programming (MINLP)-based optimization frameworks is used to select the most promising storage technology/technologies for a given power plant considering various tradeoffs between ramping capacities, ramping rates, life span and capital costs, and load balances. The framework not only helps to select the best technologies, it also provides the target capacities and the designs decisions that contribute to the overall cost of the system.

Energy generation, storage and dispatch – these are inherently dynamic operations. Specifically, a storage system needs to dynamically follow the time-varying loads and demands. The dynamic behavior must be carefully considered and modeled in order to ensure efficient power flow. Therefore, once the MINLP-based framework selects the candidate storage technologies, we consider their optimal integration with the power plant. When the system dynamics is included, it gives rise to Mixed-Integer, Nonlinear and Ordinary Differential Equation (MINODE)-based model. We propose two solution strategies to solve the MINODE problem. The first strategy is based on a time-averaged approximation, which leads to a large-scale MINLP. This is solved using a commercial solver. The second strategy utilizes a data-driven black-box optimization algorithm [5]. Both strategies are applied to illustrate the benefits of optimal downselection and integration of storage technologies for two representative regional-scale hourly electric load profiles.

References:

[1] Kumar N, Besuner P, Lefton S, Agan D, Hilleman D. *Power Plant Cycling Costs*.; 2012.

[2] Denholm P, Ela E, Kirby B, Milligan M. *Role of Energy Storage with Renewable Electricity Generation*.; 2010.

[3] Das T, Krishnan V, McCalley JD. Assessing the benefits and economics of bulk energy storage technologies in the power grid. *Appl Energy*. 2015;139:104-118.

[4] Nyamdash B, Denny E, O'Malley M. The viability of balancing wind generation with large scale

energy storage. *Energy Policy*. 2010;38(11):7200-7208.

[5] Bajaj I, Iyer SS, Hasan MMF. A trust region-based two phase algorithm for constrained black-box and grey-box optimization with infeasible initial point. *Comput Chem Eng*. 2018;116:306-321.

569041 Modular Gas-to-Liquids Process with Membrane Steam-Methane Reformer, Fischer–Tropsch Reactive Distillation and Integrated Product Refining

Shale/natural gas can be converted to liquid fuels through gas-to-liquids (GTL) processes that are applied commercially in a number of plants around the world. One of the biggest challenges associated with GTL processes is the high capital costs of these plants. Small-scale GTL have the advantages of being less capital intensive, less risk, and with the potential of utilizing gas reserves which previously could not be exploited, i.e. associated and stranded gas reserves. However, the cost per barrel of these processes still remains high, if conventional technology is employed [1]. Novel reactive distillation processes are paving the way for a more sustainable chemical process industry that is profitable, safer and less polluting [2].

Zhang et al. [3] developed an equation-oriented framework for optimal synthesis of integrated reactive distillation (RD) systems for Fischer–Tropsch processes in GAMS which has recently been implemented in PYOMO. To produce synthesis gas at the required H_2/CO ratio for the RD process, we are developing an membrane steam-methane reforming (SMR) model. This optimization-based model provides flexibility in both process design and operation. The excess hydrogen from the reformer is used to optimize gasoline and diesel production in the product refining sections.

The models are developed in Python and the optimization of the integrated flowsheet is implemented in PYOMO. The use of the SMR with internal hydrogen separation greatly improves conversion and lowers operating conditions. This type of modular design and integration of processes, which is suitable for small-scale plants, also has the potential to ease the scalability of the process.

[1] W.D. Shafer, M.K. Gnanamani, U. Graham, J. Yang, C.M. Masuku, G. Jacobs, B.H. Davis, Fischer–Tropsch: product distribution – The fingerprint of synthetic fuels, *Catalysts 9* (2019) 259.

[2] C.M. Masuku, L.T. Biegler, Recent advances in gas-to-liquids process intensification with emphasis on reactive distillation, *Curr. Opin. Chem. Eng.* (2019) https://doi.org/10.1016/j.coche.2018.12.009.

[3] Y. Zhang, C.M. Masuku, L.T. Biegler, Equation-oriented framework for optimal synthesis of integrated reactive distillation systems for the Fischer–Tropsch processes, *Energy Fuels 32* (2018) 7199–7209.

Keywords: *Equation-Oriented Optimization Framework, Detailed Kinetic Model, Small-Scale GTL, Fischer–Tropsch Synthesis, Process Intensification.*

567914 Flexible Operation of CO_2 Capture Processes at Different Plant Scales

The transition of energy systems to meet CO_2 mitigation targets has led to an increased integration of intermittent renewable energy. To balance the intermittency, fossil fuel-fired power plant with CO_2 capture and storage (CCS) will need to operate flexibly (Mac Dowell and Staffell, 2016). By coordinating the balance between electricity supply and CO_2 emission reduction targets, the economic and technical performance of power plants with CCS could potentially improve. The system-wide benefit of flexible CCS are clear, *e.g.*, total system cost is reduced, and power system resilience and operability are improved.

The profitability of the system can be maximised with flexible operation by ramping the CO_2 capture process up and down in accordance to electricity demand. In additional to this load following approach, other flexible operation strategies include systems that bypass CO_2 capture, vent flue gas, store rich/lean solvent and optimisation of the steam cycle design. Techno-economic studies demonstrate the economic value of flexible CCS operation. However, further research needs to evaluate the impact of process disturbances on the technical performance and plant operability during flexible operation. Dynamic process modelling and pilot plant studies will be essential in assessing the feasibility of flexible operation in CO_2 capture plants.

The process dynamics of a CO_2 capture plant is a function of plant size or scale. We need to better understand the effect of plant scale, also identifying the key characteristics that impact process operability and flexibility. This study will investigate the feasibility of flexible operation for post-combustion CO_2 capture at three different plant scales:

- UKCCSRC PACT CO_2 capture pilot plant (1 t_{CO2}/day capture capacity);
- Brindisi CO_2 capture plant (60 t_{CO2}/day capture capacity);
- Technology Centre Mongstad (TCM) CO_2 capture plant (80 t_{CO2}/day capture capacity).

A combination of pilot plant performance assessment and dynamic modelling have been employed to identify key factors that can limit the flexibility of a CO_2 capture plant. Dynamic models of these three CO_2 capture plants were developed in gCCS (by Process Systems Enterprise) and validated against steady state and dynamic plant data.

The UKCCSRC PACT pilot plant (Sheffield, UK) results demonstrates CO_2 capture of ~1 t_{CO2}/day from flue gas with CO_2 concentration of 12–13 vol% (similar to coal flue gas). The PACT pilot plant datasets presented in this paper demonstrates three dynamic operation scenarios: (a) partial load stripping, (ii) capture plant ramping, and (iii) reboiler steam decoupling (Bui et al., 2018). The Brindisi pilot plant (Brindisi, Italy) represents a "mid-scale" plant in this study, which captures 60 t_{CO2}/day from flue gas exiting a coal-fired power plant with a CO_2 concentration of 10–13 vol%. To study the dynamics of the Brindisi plant, step changes in the following process parameters were modelled (based on data from Enaasen et al. (2014)): steam flow rate, solvent flow rate and flue gas flow rate. The largest plant of three is the Technology Centre Mongstad (TCM) CO_2 capture plant (Mongstad, Norway), designed to capture CO_2 from two flue gas sources:

- Combined heat and power (CHP) plant: capture capacity of 80 tonnes CO_2/day from flue gas of CO_2 concentration 3.5–4 mol%.
- Residual fluid catalytic cracker (RFCC) unit: capture capacity of 275 tonnes CO_2/day from flue gas of CO_2 concentration 12–13 mol%.

The CHP configuration was used for a flexible operation test campaign at the TCM plant in July 2017. Using MEA solvent, three dynamic scenarios were implemented: (i) effect of steam flow rate, (ii) time varying solvent regeneration, and (iii) variable ramp rates.

In the development of flexible operation strategies, an important consideration is plant flexibility, which depends on: (i) the achievable minimum load, and (ii) ramp rate capabilities. Conventional combined cycle gas turbine (CCGT) power plants have ramp rates that range between 2–8 $\%P_n$/min (percentage of the nominal load per min), whereas the ramp rate of gas turbines is between 8–15 $\%P_n$ (Hentschel et al., 2016). Although the power plant ramp rate limits are relatively well understood, there is a lack reliable data on ramp rate limits for CO_2 capture processes. In comparison to large or commercial scale CO_2 capture plants, small scale plants tend to have fast dynamics (Bui et al., 2016). Therefore, the scale of a CO_2 capture plant will likely have a significant impact on achievable ramp rate limits.

A comprehensive analysis of the experimental data from three CO2 capture plants combined with dynamic model evaluation has provided invaluable insight into the process dynamics. The effect of plant scale is shown to be an important factor in the flexibility of a plant as it impacts the plant response time and feasible operating range for process parameters. Based on the pilot plant and modelling analysis, we identify the key process considerations that demonstrate the greatest influence on process dynamics. Important process parameters during flexible operation include the liquid-to-gas (L/G) ratio and lean loading (function of the reboiler temperature and L/G ratio) as both directly impact CO2 absorption performance. The factors that dictate the solvent circulation rate of the system include the solvent inventory volume, liquid flow rate, also the volume of individual equipment and pipes (determines residence time). Minimising the solvent inventory or increasing solvent flow rate can provide faster solvent circulation times. The most crucial factor is the process control system, which can limit plant flexibility (*e.g.*, restricts ramping capabilities) if not optimised for dynamic operation. The findings from this study will be essential for future design of robust flexible operation strategies and effective dynamic process control. Furthermore, we can begin to establish a correlation or function that describes the dynamic performance of a CO_2 capture process based on characteristics that describe plant scale (*e.g.*, solvent inventory, plant volume).

References

Bui, M., Gunawan, I., Verheyen, V., Feron, P. & Meuleman, E. (2016). Flexible operation of CSIRO's post-combustion CO_2 capture pilot plant at the AGL Loy Yang power station. *International Journal of Greenhouse Gas Control,* 48, Part 2 (Flexible operation of carbon capture plants), 188–203.

Bui, M., Tait, P., Lucquiaud, M. & Mac Dowell, N. (2018). Dynamic operation and modelling of amine-based CO2 capture at pilot scale. *International Journal of Greenhouse Gas Control,* 79, 134–153.

Enaasen, N., Zangrilli, L., Mangiaracina, A., Mejdell, T., Kvamsdal, H. M. & Hillestad, M. (2014). Validation of a dynamic model of the Brindisi pilot plant. *Energy Procedia,* 63, 1040-1054.

Hentschel, J., Babić, U. a. & Spliethoff, H. (2016). A parametric approach for the valuation of power plant flexibility options. *Energy Reports,* 2, 40–47.

Mac Dowell, N. & Staffell, I. (2016). The role of flexible CCS in the UK's future energy system. *International Journal of Greenhouse Gas Control,* 48, Part 2 (Flexible operation of carbon capture plants), 327–344.

566576 A Novel Approach of Enhanced Gas Recovery Using Carbon Dioxide for Shale Gas Supply Chain Network in Unconventional Reservoirs

Shale gas is one of the most important resources to improve the energy independence of the United States by supplying natural gas, and natural gas was produced and supplied locally by processing shale gas with the recent advances in horizontal drilling and hydraulic fracturing. Over the past decade, improved practices of reservoir stimulation and production have made shale gas a viable energy resource [1]. However, shale gas production by hydraulic fracturing technique includes technological and environmental issues [2]: (1) sharp decrease of shale gas production rates at a shale well; (2) increase in carbon dioxide (CO_2) emission rates by generating electricity at power plants. To achieve the economic and environmental benefits of the shale gas production, the maximum amounts of produced shale gas and reduction of CO_2 emission should be guaranteed.

Many studies have developed shale gas supply chain network (SGSCN) in an economically viable manner [2-5]; however, they did not consider handling CO_2 emission and decreased shale gas production rates after a few years of production. Enhanced gas recovery using CO_2 (EGR-CO_2), which is a technique to increase the shale gas production rates while decreasing CO_2 emissions, has gained much attention as a new approach to solve these issues simultaneously. When a huge amount of CO_2 is injected into a shale well after completing hydraulic fracturing jobs, the shale gas production rates will be increased by replacing shale gas with injected CO_2 at shale rocks. More specifically, CO_2 is injected into the shale rocks at a higher pressure than the initial pressure of shale gas in the fracture; CO_2 has a higher adsorptivity than shale gas in shale rocks, enabling it to liberate the adsorbed shale gas in place [1, 6-7]. The shale rock has the potential to store a huge amount of CO_2, which is captured from flue gas emitted at power plants generating electricity by natural gas separated from shale gas. According to these connections, EGR-CO_2 should be considered in determining optimal SGSCN configuration.

Motivated by these considerations, we focus on the development of a novel framework to apply CO_2-EGR in SGSCN model; this application will allow us to simultaneously increase shale gas production rates and decrease CO_2 emission rates. SGSCN consists of three main parts: (1) water network for ensuring the freshwater supply and wastewater treatment; (2) shale gas network for separating shale gas and generating electricity; (3) CO_2 network for ensuring the separating CO_2 and injecting CO_2 into a shale well. Based on this developed framework, the optimal SGSCN configuration will be determined by maximizing the overall profits considering EGR-CO_2 and no CO_2 injection (CO_2 emission penalty). The proposed model has been applied two case studies to illustrate its superiority over existing approaches.

Literature cited:

[1] Li, X. and D. Elsworth, Geomechanics of CO2 enhanced shale gas recovery. Journal of Natural Gas Science and Engineering, 2015. 26: p. 1607-1619.

[2] Gao, J. and F. You, Shale gas supply chain design and operations toward better economic and life cycle environmental performance: MINLP model and global optimization algorithm. ACS Sustainable Chemistry & Engineering, 2015. 3(7): p. 1282-1291.

[3] Cafaro, D.C. and I.E. Grossmann, Strategic planning, design, and development of the shale gas supply chain network. AIChE Journal, 2014. 60(6): p. 2122-2142.

Industrial & Engineering Chemistry Research 55.5 (2016): 1322-1335.

[5] Arredondo-Ramírez, Karla, José María Ponce-Ortega, and Mahmoud M. El-Halwagi. "Optimal planning and infrastructure development for shale gas production." Energy Conversion and Management 119 (2016): 91-100.

[6] Godec, M., et al., Potential for enhanced gas recovery and CO2 storage in the Marcellus Shale in the Eastern United States. International Journal of Coal Geology, 2013. 118: p. 95-104.

[7] Pei, P., et al., Shale gas reservoir treatment by a CO2-based technology. Journal of Natural Gas Science and Engineering, 2015. 26: p. 1595-1606.

575593 Rigorous, Dynamic, Two-Film Model Development and Part-Load Optimization of a Solvent-Based CO_2 Capture Process Under Variable Capture Rate

Fossil fuels will continue to play a dominant role in power generation for the foreseeable future. One potential option to reduce the carbon footprint of these plants is post-combustion CO_2 capture. However, leading post-combustion capture technologies using aqueous solvents lead to high energy penalties causing considerable decrease in plant efficiency (MacDowell et al. 2010; Wang et al. 2011). The energy penalty in these systems primarily stems from the reboiler (Freguia and Rochelle 2003; Ziaii, Rochelle, and Edgar 2009). As a result, research efforts have been focused on developing accurate predictive models to minimize the energy cost associated with solvent regeneration under various scenarios for 90% CO_2 capture (Hasan et al. 2012; Oh and Kim 2018). However, an increasing large number of power plants are cycling their load rapidly due to the ever-increasing penetration of the renewable energy sources into the grid (Roeder and Kather 2014). Thus, the plant operating conditions must be optimized under varying part-load conditions. However, tower models for solvent-based capture systems using aqueous amines are highly complex and nonlinear. Few models for these processes exist that have been validated with the part-load operational data from the pilot/industrial plants. Furthermore, rigorous models of the balance of plant (BOP) for the capture process that provides accurate estimates of their performance under part-load condition need to be developed since the BOP plays a critical role in the overall energy efficiency of the capture process. BOP models in the existing literature are generally not capable of accurate prediction under part-load operation. In this work, a detailed, dynamic tower model is developed where the mass and heat transfer between the liquid and gas phase is modeled using a two-film model. The vapor-liquid equilibrium is modeled by an electrolyte NRTL model considering the ion-ion, ion-molecule and molecule-molecule interactions, while the heat and mass fluxes are modeled using Maxwell-Stefan transport equations with reactions in the liquid phase. Maximum-likelihood estimate of parameters for the thermodynamic, kinetic, column hydraulics, mass transfer coefficient and interfacial area models is obtained using the data from contactors spanning multiple spatial scales. A rigorous algebraic model of a plate heat exchanger (PHE) based on the effectiveness-Number of Transfer Units (e-NTU) approach is developed for the lean-rich solvent cross heat exchanger, while a detailed reboiler model is presented with the chemistry model for the reactions in the liquid phase. Other auxiliary units such as the cooler, condenser, and mixer are also developed for the complete CO_2 capture process.

The plant-wide model is implemented in the modeling and optimization framework being developed by the U.S. Department of Energy's Institute for Design of Advanced Energy Systems (IDAES) (Miller et al. 2018). The IDAES framework is being built on Pyomo, an algebraic modeling language (Hart et al. 2017). The plant-wide model is validated using the operational data from the National Carbon Capture Center pilot plant under considerable variation of solvent flowrate, gas flow rate, CO_2 concentration in the gas, variable capture rate and steam flowrate to the reboiler as would be expected under the part-load operation of the host power plant. The model is used for optimizing the plant operation under part-load and variable capture rates.

References

Freguia, Stefano, and Gary T Rochelle. 2003. 'Modeling of CO_2 capture by aqueous monoethanolamine', *AIChE Journal*, 49: 1676-86.

Hart, William E, Carl D Laird, Jean-Paul Watson, David L Woodruff, Gabriel A Hackebeil, Bethany L

Nicholson, and John D Siirola. 2017. *Pyomo—Optimization Modeling in Python* (Springer).
Hasan, M. M. Faruque, Richard C. Baliban, Josephine A. Elia, and Christodoulos A. Floudas. 2012. 'Modeling, Simulation, and Optimization of Postcombustion CO_2 Capture for Variable Feed Concentration and Flow Rate. 1. Chemical Absorption and Membrane Processes', *Industrial & engineering chemistry research*, 51: 15642-64.
MacDowell, Niall, Nick Florin, Antoine Buchard, Jason Hallett, Amparo Galindo, George Jackson, Claire S Adjiman, Charlotte K Williams, Nilay Shah, and Paul Fennell. 2010. 'An overview of CO_2 capture technologies', *Energy & Environmental Science*, 3: 1645-69.
Miller, David C, John D Siirola, Deb Agarwal, Anthony P Burgard, Andrew Lee, John C Eslick, Bethany Nicholson, Carl Laird, Lorenz T Biegler, and Debangsu Bhattacharyya. 2018. 'Next Generation Multi-Scale Process Systems Engineering Framework.' in, *Computer Aided Chemical Engineering* (Elsevier).
Oh, Se-Young, and Jin-Kuk Kim. 2018. 'Operational optimization for part-load performance of amine-based post-combustion CO_2 capture processes', *Energy*, 146: 57-66.
Roeder, Volker, and Alfons Kather. 2014. 'Part Load Behaviour of Power Plants with a Retrofitted Post-combustion CO_2 Capture Process', *Energy Procedia*, 51: 207-16.
Wang, M., A. Lawal, P. Stephenson, J. Sidders, and C. Ramshaw. 2011. 'Post-combustion CO_2 capture with chemical absorption: A state-of-the-art review', *Chemical Engineering Research and Design*, 89: 1609-24.

Ziaii, Sepideh, Gary T. Rochelle, and Thomas F. Edgar. 2009. 'Dynamic Modeling to Minimize Energy Use for CO_2 Capture in Power Plants by Aqueous Monoethanolamine', *Industrial & engineering chemistry research*, 48: 6105-11.

568234 Integration of a Full-Scale Carbon Capture System with a 470-MW Lignite-FIRED POWER Plant

The team of the Energy & Environmental Research Center (EERC), Minnkota Power, Mitsubishi Heavy Industries, and Burns & McDonnell are currently performing a pre-FEED (front-end engineering and design) study for adding a full-scale carbon capture unit to Milton R. Young Unit 2 in North Dakota. When adding carbon capture to an existing coal-fired unit, the project owner generally has two options for providing steam to the carbon capture system. An auxiliary boiler could be constructed to separately generate the steam, or steam can be withdrawn from the power station's steam cycle. In the case of withdrawing steam from the steam cycle, there are several considerations that must be made to ensure that the unit will be able to function during times of carbon capture as well as when the carbon capture unit is not operating. Existing steam turbines and equipment are not necessarily designed to operate at partial steam loads over long periods of time, and the steam pressure requirements for the carbon capture system must be matched with the operational conditions of the steam cycle. Additionally, there are potential opportunities to integrate low-grade excess heat between the capture system and the power station. This paper reviews the studies that the team has performed for steam cycle and carbon capture integration at Milton R. Young Unit 2 and provides the team's early recommendations for a path forward.

562653 A Comparison of Two Bioconversion Processes for the Production of Commercially Important C_4 Streams

Four carbon compounds like 1,4-butanediol (1,4-BDO), succinic acid (SA) have been identified as platform chemicals for producing polymers from renewable resources. The successful commercialization of 1,3-propanediol by Dupont Tate & Lyle Bioproducts and 1,4-BDO by Genomatica Inc. has established the commercial viability of green chemicals. In this talk we present the techno-economic analyses to produce four-carbon compounds (C_4s) like 1,4-BDO, gamma butyrolactone (GBL), and tetrahydrofuran (THF) from dextrose through two separate routes - (i) single step bioconversion and (ii) hydrogenation of bio-based succinic acid (SA). In the single step bioconversion process for production of 1,4-BDO, capital cost of the bioconversion section was found to contribute to around 58 % of the installed equipment cost. Hydraulic load from bioconversion section and membrane filtration was found to contribute the most to operating cost. The process was found to be economically feasible with the minimum sales price (MSP) of 1,4-BDO of $1.82/kg as compared to a market price of $2.5/kg. A mechanical vapor recompression scheme used in the evaporator section was found to benefit the process, increasing energy efficiency and cost effectiveness. In comparison, the SA hydrogenation process, called the C_4s refinery, was not economically competitive given the current state of technological progress. The C_4s refinery produces 1,4-BDO, GBL, THF and ammonium sulfate (AS) fertilizer. The capital costs of the C_4s refinery were found to be 1.8 times higher while the operating costs were found to be 2.5 times higher. An important parameter is the SA fermentation titer. Increase in SA titer values to the maximum values reported in literature were not able to reduce the price of C_4s to below the market price, demonstrating the need for progress in this area. It was concluded that for an extended process involving hydrogenation of SA, additional increases in SA titer values and productivity by genetically modifying the relevant organisms will be required to achieve viability especially in comparison to the single step fermentation process for 1,4-BDO. A secondary reason for unfavorable economics was found to be the dependence of the C_4s refinery on thermal separations which contribute to significant operating and capital costs, technological progress in separation processes such as membranes for organic water separations is therefore necessary. Overall, given the published state of technology, the direct fermentation route produces C_4s (1,4-BDO) at $ 1.82/Kg and is economically viable while the C_4s refinery produces C_4s at $ 3.67/Kg and is not competitive with current production routes.

572716 An Integrated Tool for Rapid Assessment of Chemical Manufacturing Process Emissions, Treatment and Sustainability Performance

The design and assessment of chemical manufacturing processes have been aided by a variety of tools that enable the comprehensive analysis and performance testing of both the steady and transient responses of processes with different levels of complexity. However, comprehensive tools that encompass the evaluation of process emissions, treatment simulation, and sustainability assessment of chemical plant performance are yet to be presented.

In this contribution, the tools developed by the co-authors for simulating Pollution Control Units (PCUs) [1] and performing sustainability assessment with GREENSCOPE [2] are combined for the first time. The PCUs allow for the simulation of several unit operations associated with emission treatment of gas, liquid and solid streams [1], while GREENSCOPE enables sustainability analysis considering four categories: Environmental, Mass, Energy, and Economic Efficiency. Each category has several unique metrics called indicators which provide insight into the sustainability performance of any chemical process. The new integrated tool thus allows for the analysis of scenarios including the impact of current and future models of unit operations on process economics, environmental sustainability, and emissions.

This work therefore presents a unified software framework that provides the opportunity to simulate and perform emission control analysis, pollution control unit sizing and sustainability assessment that could be used to support risk evaluation for different chemical manufacturing scenarios. The developed tool is illustrated via appropriate sections of a biomass/coal gasification process for syngas production with the end goal of acetic acid manufacturing.

References

1. Li S, Feliachi Y, Agbleze S, Ruiz-Mercado GJ ,Smith RL, Meyer DE, Gonzalez MA, Lima FV. A process systems framework for rapid generation of life cycle inventories for pollution control and sustainability evaluation. Clean Technol. Envir., 20(7):1543-1561, 2018.
2. Ruiz-Mercado GJ, Gonzalez MA, Smith RL. Sustainability indicators for chemical processes: III. biodiesel case study. Ind. Eng. Chem. Res. 2013;52(20):6747–60.

577369 Exploring the Potential of Chemical Looping Systems to Provide Alternative H_2 Production Systems Solutions Using Process Simulations and Techno-Economic Analysis

The iron-based moving bed based chemical looping technology developed at the Ohio State University (OSU) is a promising candidate to yield cost benefits in natural gas utilization for a carbon-constrained scenario. OSU has developed four commercially relevant, distinct approaches for high efficiency H_2 production from natural gas[1,2,3]. The first approach focuses on a three-reactor scheme that produces CO_2 and H_2 in separate reactors, providing process intensification benefits[1]. The second approach utilizes a membrane based separation scheme to surpass the thermodynamic limits associated with the H_2 production system in the first approach[2]. The third approach retrofits the methane to syngas based co-current reaction system in a steam methane reformer and considers co-feed of steam and CO_2 for high efficiency H_2 production[3]. A fourth approach focuses on modifying the first three reactor approach and producing syngas from the reducer in addition to the Hydrogen from the oxidizer. This presentation will initially examine, how the application of these four chemical looping systems can lead to process intensification and integration opportunities. The thermodynamic performance will be further quantified by a cold-gas efficiency and an energy efficiency comparison between the three approaches and a baseline Steam Methane Reforming case with 90% Carbon capture. Relevant experiments that include high-pressure kinetic data and sub-pilot scale experimental demonstration will be used as an input for a base techno-economic analysis model for the four approaches. The integrated process and techno-economic analysis conducted with guidance from industrial partners useful and surprising insights into how chemical looping Hydrogen production approaches provide solutions under different adoption and carbon-capture scenarios.

References

1. Kathe, M. V., Empfield, A., Na, J., Blair, E., & Fan, L. S. (2016). Hydrogen production from natural gas using an iron-based chemical looping technology: thermodynamic simulations and process system analysis. *Applied Energy*, *165*, 183-201.
2. Nadgouda, S. G., Kathe, M. V., & Fan, L. S. (2017). Cold gas efficiency enhancement in a chemical looping combustion system using staged H 2 separation approach. *International Journal of Hydrogen Energy*,Â *42*(8), 4751-4763.
3. Kathe, M., Empfield, A., Sandvik, P., Fryer, C., Zhang, Y., Blair, E., & Fan, L. S. (2017). Utilization of CO_2 as a partial substitute for methane feedstock in chemical looping methaneâ €“steam redox processes for syngas production. *Energy & Environmental Science*.

570829 Economic Analysis of Battery (NMC) Cathode Material Production in Flame Spray Process with Sustainable Solvents

The cost of cathode materials contributes approximately 32.7% to the cell construction cost of lithium-ion batteries, which has a significant impact on the price of battery packs. The synthetic routes of cathode material ($LiNi_{1/3}Mn_{1/3}Co_{1/3}O_2$ (NMC333)) based on the solid-state method or co-precipitation method involve complex mixing, drying, and sintering processes, resulting in a material cost too high to reach the U.S. Department of Energy battery price target of \$125/kWh. To reduce the cathode material manufacturing cost, our lab has developed a flame spray process that utilize sustainable solvent glycerol as solvent to manufacture the cathode material. The economic analysis of the new green chemical process was studied based on discounted cash flow method. The major economic indicator used in this study is the minimum cathode material selling price (MCSP), whereas the co-precipitation method for NMC333 production was selected as the reference for economic analysis. Results show that compared with the co-precipitation method, the new flame spray process has the potential to reduce the operation cost by 31% with comparable capital investment and material cost. Moreover, MCSP of the new process is \$19.1/kg, which is 83% of that of the co-precipitation method. When the flame spray process is integrated with the sintering process inline, the MCSP is as low as 15.6 \$/kg, which demonstrates that the new process is an attractive manufacturing process for cathode material production. Furthermore, sensitivity analysis and uncertainty analysis were used to examine the effects of plant scale, raw material price, and analysis parameters on the MCSP and evaluate the certainty interval of the MCSP of the flame spray process.

574815 Soybean Oil As a Bio-Solvent to Sweeten Sour Natural Gas

Soybean oils are readily available bioresources composed of unsaturated fatty acids capable of binding sulfur compounds. Predictive models offer an opportunity to develop cost-effective methods using bio-based solvents to sweeten natural gas. While increases in fracking have increased availability of natural gas – a fuel that when combusted produces less greenhouse gases than coal or gasoline – natural gas is rich in hydrogen sulfide, rendering it sour gas. Hydrogen sulfide is corrosive to processing equipment and poses risks to human health and the environment. Predictive models such as COSMO-RS, the conductor-like screening model for real solvents, can be used to examine interactions at a molecular level, screen solvents, and choose process parameters. COSMO-RS was used to simulate the partitioning of hydrogen sulfide between soybean oil, high oleic soybean oil, free fatty acids, and gas phases. Predicted partition coefficient (K) values ranged from 0.16 to 0.70 for soybean and high oleic soybean oils at temperatures from 10 - 100°C; K values between 0 – 1 are indicative of hydrogen sulfide's preference for the liquid phase over the gas phase. Experimental results demonstrate the feasibility of removing H_2S from gas using soybean oils and are compared to the predicted partitioning. This study provides proof-of-concept for using high oleic soybean oil as a bio-based solvent as a new method for cleaning sour gas.

574828 Reducing Energy, Solvent, and Resin Demand of Sorption-Based Processes for Carboxylic Acid Recovery

Current methods such as ion exchange and distillation to separate bio-based carboxylic acids are material and energy intensive, typically generating waste streams and requiring energy intensive distillation steps to recover a pure product. Sorption-based approaches increase the sustainability of carboxylic acid separations by reducing downstream waste handling, however, current sorption-based technologies typically require large volumes of elution solvent and still result in a dilute carboxylic acid stream. Thus, to obtain a concentrated product, excess water and solvent must be removed from the elution stream via evaporation or distillation which adds a substantial operational cost to the process. This presentation will discuss recent work towards reducing solvent, adsorbent requirements and reducing downstream distillation energy demands in these processes. Specifically, four commercially available weak-base resins were screened by determining their adsorption capacity for butyric acid, a fermentable acid targeted for biofuels. Then, to recover the adsorbed acid, non-aqueous solvents were tested and their regeneration constant were determined. Findings indicate that specific properties of the resins and eluents correlate with the energy, resin, and solvent demands of the system. These results provide a general framework for optimizing sorption-based processes for recovering dilute bioproducts.

572591 Storage of Excess Wind Energy and Utilization of Waste Flue Gas CO_2 through Sustainable Methanol Production

As a larger production of wind energy and other types of renewable energies are predicted in the future, utilizing the excess energy from these sustainable energy sources can increase the energy efficiency of the wind farm by making good use of all the electricity produced. Moreover, CO2 emissions from industry, that contribute to climate change, are also an increasing problem that requires a solution. In this study, we present a novel concept, which aims at solving these two problems by integrating excess energy from wind farms and CO2 from flue gas into an innovative production process using methanol as an energy storage/carrier. The excess wind energy is used to produce hydrogen through water electrolysis and CO2 is captured from flue gas, e.g. from the cement industry. Compared with the traditional methanol production which heavily relies on natural gas as its source of hydrogen, carbon, and energy, this process has a smaller environmental footprint. The novelty of our concept relies on shifting the focus from operating a methanol production plant towards valorizing the excess renewable energy, which is wasted otherwise and considering methanol as a value-added product (as opposed to Szima et al. 2018). Therefore, the plant capacity of 30 000 metric tons per year is scaled based on the quantity of the excess wind energy of the largest wind farm in Denmark. This capacity is significantly lower than those of traditional methanol production plants, which range up to 1.9 million metric tons per year (Ferrostaal, 2015). This adds a constraint to improve the process integration efficiency of the proposed concept. Hence, a techno-economic analysis is performed to investigate the overall feasibility of the proposed concept by identifying key design decisions and essential cost drivers. As results obtained show high sensitivity to both electricity and methanol prices, an uncertainty analysis was conducted in order to determine their effects on the profitability metrics of the proposed novel process. A sustainability analysis is complementarily carried out to emphasize the added environmental benefits of the proposed concept in comparison to the traditional methanol production process.

Reference:

Szima, Szabolcs, and Calin-Cristian Cormos. "Improving methanol synthesis from carbon-free H2 and captured CO2: A techno-economic and environmental evaluation." *Journal of CO2 Utilization* 24 (2018): 555-563.

Ferrostaal. "M5000 Methanol Plant" from *www.ferrostaal.com/en/top-references-ferrostaal/m5000-methanol-plant/* (2015), visited 12.04.2019.

575320 Towards a Novel Energy Financial Security: The Texas A & M Energy Spot Price Index

The available financial instruments providing exposure for the investors to the energy markets are limited to equity based stocks (i.e. S&P 500 Energy) and futures on individual energy commodities (i.e. WTI Crude Oil, Henry Hub Natural Gas). Furthermore, these options are predominantly associated with the oil and gas sector, with almost zero emphasis placed on the renewable energy. Given the lack of a financial security to effectively capture the entire U.S. energy landscape and represent the average cost of energy, we introduce the Texas A&M Energy Spot Price Index (ESPI) as a novel framework to fulfill these needs.

The energy landscape of the U.S. is carefully depicted to determine the products that are directed to the end-use sectors of the U.S. economy [1-5]. The total energy demand of these products, together with their spot prices in the market, serve as the backbone of the Energy Spot Price Index. Since the weights of the Energy Spot Price Index constituents have a lag between two and three months, we introduce a rolling horizon based parameter estimation model to determine the current weights. The spot prices of the products are taken from Bloomberg when available. For products whose spot prices are not available, we introduce and examine three different approaches [6-8]. The methodology is illustrated on 120 months to show the capability of accurately determining the value of the Energy Spot Price Index. The proposed Energy Spot Price Index is the first of its kind that precisely quantifies the average market price of energy. As such, it can be utilized as an efficient tool for all investors who look for exposure over the average market price of energy, trading over the whole energy market, hedging their investments, as well as a benchmark for measuring the overall cost of energy. Therefore, the Energy Spot Price Index can be used as a tradable financial security (e.g. ETF) that is updated daily providing an alternative for the investors who look to diversify their portfolios.

References:

- Elia, J. A.; Baliban, R. C.; Xiao, X.; Floudas, C. A. Optimal energy supply network determination and life cycle analysis for hybrid coal, biomass and natural gas to liquid (CBGTL) plants using carbon-based hydrogen production. Computers and Chemical Engineering 2011, 35, 1399-1430.

- Elia, J. A.; Baliban, R. C.; Floudas, C. A. Nationwide energy supply chain analysis for hybrid feedstock processes with significant CO2 emissions reduction. AIChE Journal 2012, 58 (7), 2142-2154.

- Elia, J. A.; Floudas, C. A. Energy Supply Chain Optimization of Hybrid Feedstock Processes: A Review. Annual Review of Chemical and Biomolecular Engineering 2014, 5 (1), 147-179.

- Onel, O.; Niziolek, A. M.; Floudas, C. A. Integrated biomass and fossil fuel systems towards the production of fuels and chemicals: state of the art approaches and future challenges. Current Opinion in Chemical Engineering 2015, 9, 66-74.

- Floudas, C.A.; Niziolek, A. M.; Onel, O.; Matthews, L.R. Multi-Scale Systems Engineering for Energy and the Environment: Challenges and Opportunities. AIChE Journal 2016, 62 (3), 602-623.

- The Texas A&M Energy Price Index: A Predictive Framework, 2019

- Energy Information Administration (EIA), Electric Power Monthly, 2019

- U.S. Department of Energy (DOE), Green Power Markets, 2018

566273 Techno-Economic Analysis of Septic Tank Waste Treatment in a Centralized Hydrothermal Treatment Facility

Hydrothermal treatment, a thermochemical conversion process that is particularly useful for wet wastes, of septage produces valuable solid and liquid fertilizers while eliminating human pathogens. Additionally, it treats a problematic waste that is a burden for most traditional waste water treatment facilities. Operation of a centralized treatment facility would augment an area's capacity to utilize the nutrients and valuable products available in septage while also mitigating its potential environmental hazards. This facility would be alternative to landfilling, which can be detrimental to public waterways, or treatment at a traditional waste water plant. This analysis investigates the most important factors for an economically feasible facility that both treats septic tank waste and produces fertilizer products. The environmental impact of the plant is also compared to the standard methods of septage treatment.

578172 Turning a Hazard into Sustainable Revenue Streams: H_2S Valorization in Offshore Oil and Gas Productions through Process Intensification

The grand challenges of the 21st century lie squarely at the interplay of water, energy, and food— the so-called water-energy-food nexus. Add climate change into the mix, and the challenges become *more* challenging. Aggressive measures to decarbonize the energy market are necessary to tackle the threat of climate change. However, due to the ever-increasing global population, energy requirements are expected to increase 50% by 2030.[i] While ensuring that we protect the natural environment, it is also important to provide energy security. This calls for an energy supply that is consistent, reliable, sufficient, and affordable. Natural gas has arguably the least carbon footprint among fossil energy resources (carbon dioxide emission of burning natural gas is approximately half that of coal). While natural gas cannot be viewed as the permanent solution to energy security, it plays a vital role as the bridge fuel in the transition to a low-carbon economy.

Hydrogen sulfide (H_2S) is a major hazardous by-product in oil and gas production as well as many bio-based manufacturing processes, such as anaerobic digestion. Desulfurization of sour gas is essential prior to transportation or further processing due to its extremely toxic, acidic, and corrosive nature. Several offshore oil and gas production platforms are suffering from elevated levels of H_2S in recent years, and the production facilities are not designed for such high levels of H_2S. Currently, H_2S is removed using a non-regenerative scavenging process. The scavenger reacts with H_2S and produces less acidic, water-soluble compounds in an irreversible reaction.

Although the current process is rather established, it comes with significant challenges. Such challenges include high OPEX due to excessive amounts of the scavenger used in offshore productions, environmental concerns coming from the increasing levels of H_2S and increased mass of discharged spent scavenger, and fouling issues creating blockages in several downstream unit operations, such as wastewater treatment facilities.

For these reasons, design and development of more sustainable approaches that can fulfil the H_2S content threshold limit requirements and that can eliminate the disadvantages of the current non-regenerative H_2S-scavenging process are deemed necessary, specifically from an economic point of view. This work focuses on the design and techno-economic analysis of an intensified process that can be installed on an offshore platform to simultaneously remove and convert H_2S from produced natural gas to value-added product(s) to generate new revenue streams. The added benefits of this new process are that it allows for the efficient recovery of resources by turning the hazard, H_2S, into an opportunity. The proposed process is therefore well suited for the application of intensification concepts, integrating reaction and separation into a single unit.

[i] IEA, "World Energy Outlook 2012," International Energy Agency, 2012.

554760 Critical CO_2 Carnot Engine for Industrial Waste Heat Recovery and Utilization

Industrial processes regularly discard large volumes of low-grade waste heat to the environment. Harnessing this heat and transforming it into usable power would be environmentally and economically advantageous for industries. An overlooked solution to utilizing this waste heat is the use of a Carnot cycle to generate work with supercritical CO_2 as the working fluid. Supercritical $CO_2$is a novel working fluid for a Carnot engine as its critical temperature (31.1C) is much lower than the boiling point of water. This allows for the lower isotherm of the Carnot cycle to be constructed at the critical temperature, resulting in a sizable Carnot cycle with a near flat lower isotherm and upper isotherm just below the temperature of the low-grade heat. Thus, this type of Carnot cycle encompasses a larger area on a P-V diagram than any other thermodynamic conditions, thereby optimizing the amount of work that can be converted to power using a relatively small engine.We describe this approach of waste heat harnessing and illustrate ideas for utilizing the waste heat from an industrial toluene process. The impact of compression ratio, upper isotherm temperature, and pressure in the engine are investigated. Resulting savings amount to hundreds of thousands of kWh/year, thereby reducing cooler loads and electricity bills for the process, as well as reducing environmental emissions.

References:

[1] J.T. Banchero, B.D. Smith (Ed.), R.J. Hengstebeck. *Disproportionation of toluene*, 1969.

[2] Eldred H. Chimowitz Madeleine R. Laitz, F. Douglas Kelley. *Critical CO_2Carnot Cycle for Waste Heat Utilization*, 2017.

579288 Recovery of Rare Earth Elements and Critical Materials from Coal and Coal-Based Resources

DOE-NETL's *Feasibility of Recovering Rare Earth Elements* Program has demonstrated the capability of generating rare earth elements (REEs) from coal and coal-based resources. R&D efforts are now focused on extracting, separating, and recovering REEs in domestic, bench- and small pilot-scale, coal-based, REE-processing systems. An overview of DOE-NETL's REE Program and its technical accomplishments is presented.

587890 High Yield and Economical Production of Rare Earth Elements from Coal Ash

Coal and coal byproducts are a potential source of rare earth elements (REEs), with particularly heavy rare earth (HREE) content. Demonstration of technologies to recover REE, yttrium and scandium (REYSc) from these feedstocks will enable utilization of coal mining/coal combustion waste products in environmentally benign ways to produce strategically important materials for the United States while generating jobs and economic growth. Under sponsorship from the Department of Energy, the team of Physical Sciences Inc., University of Kentucky/Center for Applied Energy Research, and Winner Water Services is developing and demonstrating a pilot scale plant to economically produce salable REYSc-rich concentrates and commercially viable co-products from coal ash feedstock using environmentally safe and high-yield physical and chemical enrichment/recovery processes. A recently commissioned pilot plant operates at the scale of approximately 0.4-1 tons per day (tpd) ash throughput for physical processing and about 0.5 tpd for chemical processing, producing at least 50 - 500 g of dry REYSc nitrates concentrate containing at least 10-20 wt.% REYSc (elemental basis). Various ash feedstocks have been qualified, and all contain between 400-500 ppm of REYSc. The data obtained from the pilot plant operations will be used to enhance and validate a commercial scale techno-economic analysis that was completed for both the physical and chemical processing plants. Pilot plant processing parameters will also be used to design a commercial scale plant (hundreds of tpd throughput) with return on investment in less than seven years. This presentation will describe the approach and the key results with regard to process optimization, scale up and pilot plant implementation. The talk will provide a comprehensive overview of the overall process for economical production of REY concentrates. The production of valuable co-products will also be discussed. Development and demonstration of modular pilot scale physical and chemical plants for REYSc recovery from coal ash, including enhancement and validation of a techno-economic model using pilot plant operations data, will enable the design of a commercial scale (hundreds of tpd) REYSc plant.

Acknowledgement: This material is based upon work supported by the U.S. Department of Energy under Award DE-FE0027167.

574967 Identification of Optimum Coal Byproducts for Recovery of Rare Earths

Abstract

The United States has the world's largest coal reserves. The lanthanides can be valuable products from abundant domestic coal and coal byproducts. There are numerous coal and coal byproduct resources, including coals, bottom ash, fly ash, coal preparation materials, miming wastes, and stored or ponded ashes. The characteristics that define the best resources, including availability; abundance; concentration of the rare earths and other critical elements; ease of extraction; production of co-products; and co-benefits such as environmental remediation, location, and transportation will be outlined, and a first attempt at identification of the best domestic coal and byproducts for recovery of the lanthanides will be made.

568318 Effect of Acidity and Particle Size on the Kinetics of Leaching of Rare Earth Elements from Coal Ash

Rare earth elements (REEs) are comprised of the 15 lanthanide elements, plus yttrium, and scandium. The objective of this study was to determine the kinetics of extraction of REEs from various fly ash, ponded ash, bottom ash, and landfill ash samples. Both strong and weak acids were utilized for the batch extractions. An expression for predicting rare earth dissolution rates based on acidity and particle size was derived.

572826 Utilizing Biooxidation in a Flow Sheet to Facilitate Low-Cost Rare Earth Elements Extraction from Coal-Based Resources and Eliminate Future Acid Mine Drainage

Michael L. Free, Prashant K. Sarswat, Landon Allen, Aaron Noble, and Gerald Luttrell

Coal-based resources such as waste rock near the edges of coal seams often contain significant quantities of rare earth elements. These resources often end up on waste piles. In many locations, the waste piles become acidic after long periods of time, and the associated acidic environment causes a slow release of metals, including some rare earth elements (REE) that are lost in waste treatment facilities. These resources can be recovered economically through appropriate processing of the coal feed material to produce a waste that is concentrated in rare earth elements. This process can be facilitated using heap leaching with solutions from engineered biooxidation of pyrite naturally found with the coal resource, and the leached REEs subsequently recovered through solvent extraction and precipitation. In order to address this opportunity for low cost REE extraction and recovery, the objectives of this study and the related DOE-funded project are to demonstrate and improve methods to economically extract, recover, and upgrade the rare earth element (REE) contents from vast coal-based resources using integrated modeling, targeted coal preparation, biooxidation, solution conditioning, heap leaching, solvent extraction, and precipitation technologies to produce REE-bearing products in an environmentally friendly way. Utilization of these integrated technologies will enable coal producers to utilize untapped resources to produce revenue and extend resource life while simultaneously reducing future environmental issues and costs as they supply critical domestic REEs. Although this approach will be designed for coal ore processing facilities, it will be easily adapted in the future to the recovery of REEs and mitigation of acid rock drainage from coal tailings. Thus, this technology will enable recovery of an underutilized resource in an environmentally friendly way. Initial results from this study will be presented and discussed.

573532 Economic Extraction and Recovery of Rees and Production of Clean Value-Added Products from Low-Rank Coal Fly Ash

The University of North Dakota Energy & Environmental Research Center (EERC), with assistance from Pacific Northwest National Laboratory and the U.S. Department of Energy's National Energy Technology Laboratory, is developing economically viable and tailorable rare-earth element (REE) extraction and concentration methods for low-rank coal fly ash. These methods are tailored based on a detailed understanding of fly ash formation mechanisms, fly ash properties, the form of the REEs in the fly ash and, to some degree, the mode of occurrence of the REEs in the coal. These new methods were applied to high-calcium-containing fly ash materials derived from the combustion of subbituminous coals from the Powder River Basin (PRB) and North Dakota lignite coal from the Fort Union region, the resource of which has the potential to far exceed the current U.S. REE demand. The EERC evaluated several simple and environmentally benign pretreatment approaches that can 1) alter the physical structure of the fly ash by increasing porosity or decreasing particle size; 2) alter the chemical composition of the fly ash to improve solubility, allowing the use of a much milder leaching process to extract REEs; and 3) enrich the metals/elements of interest on the surface of the ash particles, making them more accessible by solvents. This approach was used to take advantage of the high alkali/alkaline-earth content in PRB and lignite fly ashes—a significant REE recovery advantage over eastern bituminous coals. In addition, the EERC will present results from the development of improvements to existing supercritical CO_2-based extraction of metals from ash and other solid waste materials. This involves a lower-cost method of compressed CO_2 processing that utilizes both traditional and novel, inexpensive metal chelating agents.

571678 Recovering Rare Earth Elements from Acid Mine Drainage Using Coal Combustion By-Products through Mine Reclamation

The demand of rare earth elements (REEs, including scandium, yttrium and a group of 15 lanthanides) in making critical parts for emerging technologies is projected to grow over the coming decades. Due to the significant instability of the market, finding alternative sources has become a critical issue for the United States and other countries. In the U.S., acid mine drainage (AMD) is considered a potential alternative source for REEs. Currently, the research team is developing a two-stage trap-and-concentrate (TAC) process to prepare a feedstock for producing rare earth oxides (REO feedstock). The TAC process first uses stabilized flue gas desulfurization material (sFGD) to capture REEs from AMD (AMD REEs) and then applies a sequential extraction procedure to produce a REO feedstock containing above 2 wt.% of total rare earth elements (T-REEe). sFGD material is a mixture of lime (CaO), calcium sulfite FGD by-product, and coal ash. The TAC process can be integrated with abandoned mine land (AML) reclamation to create an approach that can add economic incentives for AML reclamation, remediate AMD discharge, provide a long-term, high-volume beneficial use for coal combustion by-products, which otherwise needs to be disposed of in a landfill, and eliminate public safety hazards and threats to local environment and ecological systems posed by AMLs. The objectives of this DOE project are to (1) validate and understand the TAC process, (2) quantify the associated economic and environmental benefits, and (3) evaluate the full-scale application potential. So far, we have demonstrated that the sFGD material is effective in trapping AMD REEs. Over 99% of AMD T-REEe in contacted with sFGD material can be retained under a wide range of reaction conditions. A number sequential extraction approaches are tested to concentrate the retained REEs. The factors controlling the recovery of AMD REEs in the TAC process is investigated by using synchrotron-based analytical techniques to explore molecular-level understandings on the involved chemical mechanisms. The economic and environmental implications and the benefits of combining the TAC process with AMD mitigation as an integrated AML reclamation approach are also evaluated.

575745 Membrane and Sorbent Based Recovery of Rare-Earth-Elements (REE) from Acid Mine Drainage Sludge

This project presents results from a staged membrane-based and sorbent based, bench-scale system to extract rare earth elements (REEs) from acid mine drainage (AMD) sludge generated as part of coal mining activities in the United States. The treatment approach includes a review of feedstock characterization and results from separation, concentration, and ultimate recovery of REEs from AMD. Sludge samples from four AMD site(s) were characterized for REE concentration, dissolved metals concentration, and key water-quality characteristics. Each individual process component were tested with water samples to optimize performance. The work demonstrates proof-of-concept experiments at the bench scale with respect process parameters such as water chemistry, nanofiltration membrane performance in monovalent/multivalent separation, affinity media chemistry on the final recovery of REE.

579283 Developing Low-Cost, Efficient, Gasification-Based Modular Energy Systems

This paper provides an overview of the U.S. Department of Energy's National Energy Technology Laboratory (DOE/NETL) Gasification Program's work to develop smaller-scale, low-cost, and efficient coal gasification-based energy systems. Research and development is focusing on implementing coal gasification processes into small modular systems having high efficiency for power generation or combined heat and power. These systems would rely on process and/or reaction intensification, innovative oxygen separation technologies with superior performance at smaller scales, and other enabling technologies. Accelerated technological development is possible using NETL's sophisticated modeling and simulation capabilities; advanced materials and fabrication methods; and by leveraging NETL's partnerships with industry, academia, and other national laboratories.

570993 Upgrading Coal for the Production of Value-Added Chemicals and Feedstocks By Wave Liquefaction™: Spectroscopic Diagnostics and Material Characterization

1.0 Introduction

Coal liquids and tars are valuable chemicals with applications ranging from metallurgy, energy storage, wood preservatives and carbon fibers [1]. Coal pyrolysis, a thermochemical conversion of coal in oxygen deficient environment at high temperature produces liquid (coal-tar), solid (coal-char) and gases [2]. The main chemicals present in the coal-tar are benzene, toluene, xylene and phenolics, which serve as valuable feedstocks for the petrochemical industry [3]. In conventional pyrolysis the solid particles are heated to desired pyrolysis conditions by the external heating, e.g. electrical heating, hot sand or process steam [2]. These conventional heating strategies suffer from heat transfer limitations, which reduces the overall process energy efficiency and product yield [4]. Liquid, tar yields and their product distributions depend upon heating rate, generally increasing with temperature (up to a limit) [5]. Microwave (MW) driven pyrolysis is a promising approach to resolve these limitations. Energy efficiency is a dramatic advantage of microwave heating: the typical energy efficiency of a kitchen microwave oven magnetron at 1kW output power is around 65-70%, while industrial scale magnetrons demonstrate about 80% energy efficiency at output power on the order of 100 kW [6]. A key innovation is the ability to process coal with little to no CO_2 emissions or water consumption. In the literature to-date, microwave pyrolysis of virgin coal has been largely applied to other coal processing purposes, rather than devolatilization.

In reactive atmospheres (H_2, CH_4 or both) microwave radiation forms a plasma with energetic electrons creating H-atoms and methyl radicals [7]. These reactive species can cap radicals, add to -bonds and initiate bond scission [8]. Radical capping through hydrogenation or methylation (in H_2/CH_4 reactive atmosphere), suppresses both retrogressive and secondary cracking reactions [9] that are typical of the conventional pyrolysis processes, decreasing liquid and tar yields [10]. In fact radical termination is essential to liberating liquids and tars as labile groups depart and matrix linkages scission, preventing charring [11]. Such capping reactions by H-atoms and methyl radicals also upgrade unsaturated and aromatic compounds by increasing the H/C ratio. Moreover these reactive species can potentially promote coal matrix deconstruction by radical abstraction/addition reactions, initiating bond scission.

This paper presents optical spectroscopic diagnostics as used in an engineering pilot implementation of H Quest Vanguard's microwave plasma process applied to coal, and material characterization results from application of the process to H_2 generation from natural gas. Plasma diagnostics for process control will be illustrated. Chemical analyses are presented of coal conversion products. Recent results yielding nanographene as a premium carbon by microwave-plasma processing of methane will be shown.

2.0 Measurements and Analyses

Here a suite of characterization tools have been applied to the light gases, liquid and solid carbon products. Gas and liquid analyses were performed using gas chromatography-mass spectrometry. For the solid carbon products SEM has been applied to coal chars for cenosphere analysis, and for the carbon products from methane decomposition, scanning electron microsocpy (SEM) for morphology, transmission electron microscopy (TEM) for (nano)structural assessment and identification of different carbon forms (i.e. sp^2 phases), X-ray diffraction (XRD) for evaluation of graphitic structure, thermo-gravimetric analysis (TGA) for bulk determination of oxidative reactivity of the carbons as a means by which to assess graphitic content and gas (N_2) adsorption analyses for texture – i.e. surface area and porosity. Elemental analysis was also performed for HCNS and O by difference. Complementary thermo-gravimetric analyses assess the oxidative reactivity of the carbon products. Optical diagnostics such as multi-wavelength pyrometry relate these material characteristics to reactor conditions and can be applied for process control.

Correspondingly optical diagnostics are central to reaction characterization and hold particular value for species identification and temperature determination [12]. In this study spectroscopic data collection was performed using a fiber-optically coupled Ocean Optics HR2000 spectrometer fitted with a UV-NIR grating for a spectral range of 900 nm. All optical emission spectra were captured with equipment spectrally calibrated using a Hg/Ar lamp and corrected for instrument response function using a NIST-traceable tungsten lamp as spectral calibration standard. Temperatures of C_2* are determined using custom spectral band-fitting algorithms, taking into account the electronic baseline and underlying blackbody radiation. Spectral fits are performed at 1 nm spectral resolution. Carbon particle temperatures are determined by fitting Planck's radiation law with black body conditions.

Observed intensity ratios or spectra band shapes can yield temperature by Boltzman analysis using spectral constants. Moreover, optical emission serves to identify reactive species and intermediates, albeit indirectly inferred by the observation of their electronically excited counterparts, e.g. CH* and C_2* radicals. For example, the presence of key atomic or diatomic radicals can support postulates of electron impact dissociation, and radical mediated bond insertion or radical capping reactions and provide mechanistic insights from the temperatures associated with the different degrees of freedom – electronic, vibrational, such as from the C_2* ($d^3{}_g - a^3{}_u$) Swan band emission.

3.0 Summary

H Quest Vanguard is developing broad-spectrum microwave plasma processes targeting conversion of hydrocarbon feedstocks such as coal and natural gas to value-added materials, chemicals and fuels [13,14]. A microwave pyrolysis approach was originally proposed in response to the DARPA initiative for green-house gas (GHG)-emission-free and cost-effective production of US Air Force jet fuel from the domestic coal resources. In H Quest Vanguard's process, natural gas can be used in single-stage reactor as a hydrogen source, eliminating external hydrogen production units and the associated CO_2 production, water consumption, and capital costs, and providing excess hydrogen sufficient for downstream hydro-treating. The MW process produces fast heating resulting in flash devolatilization and pyrolysis followed by fast quenching, preserving the primary pyrolysis products and volatiles' molecular structure.

Since company formation in 2014, H Quest Vanguard, Inc. has developed ample material base and expertise in development of chemical and catalytic processes enhanced by microwave plasma. In particular, H Quest has developed a conversion process that applies microwave energy to rapidly co-pyrolyze solid hydrocarbons (e.g. coal) and natural gas to produce liquid hydrocarbons. A wide range of carbon materials including graphene and ordered carbon blacks have been observed across a wide range of experiments. These forms have potential for high value applications: electrical conductivity additives for plastics, and as electrode material in supercapacitors and Li-batteries. Subsequent and ongoing work has demonstrated feasibility of a microwave driven plasma mediated methane decompostion for H_2 production and nanographene formation as a premium carbon.

4.0 Acknowledgements

H Quest Vanguard, Inc. is a privately held technology company, based in Pittsburgh, Pennsylvania, focused on the development and commercialization of novel hydrocarbon conversion technologies. This material is based on work supported by the American Chemical Society, Petroleum Research Foundation, Award No. PRF# 58973-ND4 and the Department of Energy, Office of Science through subaward agreement no. 186949 with H Quest Vanguard, Inc. under the Prime Award DE-SC0015895 Phase I SBIR.

5.0 References

1. Schlosberg, R. H. (2013). *Chemistry of coal conversion*. Springer Science & Business Media.
2. Ladner, W. R. (1988). The products of coal pyrolysis: properties, conversion and reactivity. *Fuel processing technology*, *20*, 207-222.
3. Schobert, H. H., Song, C. (2002). Chemicals and materials from coal in the 21st century. *Fuel*, *81*(1), 15-32.

4. Bridgwater, A. V. (2012). Review of fast pyrolysis of biomass and product upgrading. *Biomass and bioenergy*, *38*, 68-94.
5. Yan, B. H., Cao, C. X., Cheng, Y., Jin, Y., Cheng, Y. (2014). Experimental investigation on coal devolatilization at high temperatures with different heating rates. *Fuel*, *117*, 1215-1222.
6. Bermúdez, J. M., Beneroso, D., Rey-Raap, N., Arenillas, A., Menéndez, J. A. (2015). Energy consumption estimation in the scaling-up of microwave heating processes. *Chemical Engineering and Processing: Process Intensification*, *95*, 1-8.
7. Wang, N., Yu, J., Tahmasebi, A., Han, Y., Lucas, J., Wall, T., Jiang, Y. (2013). Experimental study on microwave pyrolysis of an Indonesian low-rank coal. *Energy & fuels*, *28*(1), 254-263.
8. Şimşek, E. H., Karaduman, A., Olcay, A. (2001). Liquefaction of Turkish coals in tetralin with microwaves. *Fuel Processing Technology*, *73*(2), 111-125.
9. Solomon, P. R., Serio, M. A., Suuberg, E. M. (1992). Coal pyrolysis: experiments, kinetic rates and mechanisms. *Progress in Energy and Combustion Science*, *18*(2), 133-220.
10. Durai-Swamy, K. (1982). *U.S. Patent No. 4,324,642*. Washington, DC: U.S. Patent and Trademark Office.
11. Retcofsky, H. L., Hough, M. R., Maguire, M. M., Clarkson, R. B. (1981). Nature of the free radicals in coals, pyrolyzed coals, solvent-refined coal, and coal liquefaction products.
12. Vander Wal, R. L., Gaddam, C. K., and Kulis, M. J. (2014). An Investigation of Micro-Hollow Cathode Glow Discharge Generated Optical Emission Spectroscopy for Hydrocarbon Detection and Differentiation. *Applied spectroscopy*, *68*(6), 649-656.
13. Skoptsov, George L., and Alan A. Johnson. *Method for processing hydrocarbon fuels using microwave* energy. U.S. Patent 9,682,359, issued June 20, 2017.
14. Strohm, J.J., Linehan, J.C., Roberts, B.Q., McMakin, D.L., Sheen, D.M., Griffin, J.W. and Franz, J.A., Battelle Memorial Institute, 2012. *Heavy Fossil Hydrocarbon Conversion and Upgrading Using Radio-Frequency or Microwave Energy*. U.S. Patent Application 13/401,216.

577490 Analysis of the Heat Transfer in a Packed Bed Composed of Alumina and Coal Exposed to a High Thermal Radiation Flux

ANALYSIS OF THE HEAT TRANSFER IN A PACKED BED COMPOSED OF ALUMINA AND COAL EXPOSED TO A HIGH THERMAL RADIATION FLUX

Summary

This work investigates temperature profiles and heat fluxes in a packed bed composed of alumina and coal particles heated by radiation that simulates solar energy flux with the purpose of hydrogen production. For this purpose, the discrete element method (DEM) was used along a heat and mass transfer model. As results of the simulations, thermal profiles through the hybrid packed bed were obtained, as well as the accompanying heat flux and hydrogen production. The three chemical kinetics models tested showed good correlations between each other. It is concluded that the radiation and chemical kinetics model proposed provides the basis to continue with the experimental study of granular materials at high temperature, where radiation is the dominant mechanism.

Key-words: Hybrid packed bed; Coal gasification; Discrete element method (DEM); Radiation; Heat transfer

---

556742 Conversion of Coal Wastes and Municipal Solids Mixtures By Pyrolysis Torrefaction and Entrained Flow Gasification

ease energy independence and security while minimizing or eliminated the waste stream associated with coal mine and processing facilities. Reaction pathways for the conversion of coal wastes and coal plus opportunity fuel mixtures for modular reactive systems were experimentally explored. A bench-scale batch torrefaction reactor used for torrefaction of the municipal solid waste (MSW) to create a coal-like feed for blending with the coal waste feedstock. The conversion kinetics of the MSW feedstock were analyzed through a thermogravimetric analysis (TGA) to determine optimal process conditions for the conversion system. Torrefied MSW/biomass and waste coal blends were processed in a pilot-scale entrained flow gasifier (EFG) equipped with emissions monitors for H_2, CO, CH_4, CO_2, and H_2S together with gas chromatograph equipment for measuring syngas compositions. Product yield and emissions with coal waste plus municipal solids (CWPMS) fuels as well as the ash content and slagging properties of these feedstocks were measured. The PT-EFG system was experimentally validated by the pilot-scale gasifier, where emissions and yield data were used to develop a high-fidelity process model and economic analysis. TGA experiments were used to determine the optimal process conditions for the torrefaction of the MSW, such as reaction temperature, which preserves the energy yield of the fuel while maximizing the mass yield of the feedstock. Using this data, a bench-scale torrefaction reactor was designed and tested to further examine the yields of the torrefied feedstock in larger batch processes. Reactor temperature was verified through the bench-scale experiments and a large batch of material was torrefied and prepared for testing in the pilot-scale EFG. Detailed ash analysis and ash viscosity calculations were conducted to prove the feasibility of processing these feedstocks in an EFG, which are typically used for low ash content, low-ash viscosity feedstocks.

576367 Moving Bed Chemical Looping Reactor Systems for Power Generation and Hydrogen – Pilot Scale Testing and Process Analysis

Researchers are developing chemical looping technologies to convert carbonaceous fuels to high value chemicals and/or electricity with minimal CO_2 emission. These processes use a metal oxide or metal sulphate to partially or fully oxidize the fuel source to a desired product while being regenerated with air and/or steam in a separate reactor. The chemical looping redox reaction pathway is capable of high product yields without the need for molecular oxygen and minimizes gas product separation requirements. The Ohio State University (OSU) has advanced the chemical looping concept in the development of 2 pilot-scale demonstration plants for CO_2 capture applied to coal combustion for power production and gaseous fuel conversion to high purity H_2 as well as multiple sub-pilot test units for syngas generation from biomass, natural gas, and coal. In each of these processes, a moving bed reducer reactor is used for partial or full conversion of the carbonaceous fuels to CO/H_2 and CO_2/H_2O, respectively. A fluidized bed combustor reactor is used to regenerate the oxygen carriers to their original oxidation state. Interconnecting nonmechanical valves between reactors are used to control the solids circulation rate of the system for the mass balance of the redox reaction and heat balance for the system integration. The present summarizes the key developments of four chemical looping processes – the 250kW$_{th}$ syngas chemical looping (SCL) pilot plant, the coal direct chemical looping (CDCL) 25 kW$_{th}$ sub-pilot unit and 250 kW$_{th}$ pilot plant, and the 15 kW$_{th}$ coal to syngas (CTS) and shale gas to syngas (STS) sub-pilot unit. Key process design features and experimental results from over 1,500 hours of cumulative operation will be discussed. The counter-current moving bed reducer reactor design the CDCL and SCL process ensure nearly full fuel conversion to CO_2 with minimal solid circulation and capability of producing high purity H_2. The co-current moving bed reducer reactor design in the CTS and STS processes provides a desirable gas-solid contacting pattern that minimizes carbon deposition and maximizes the syngas yield. The syngas produced by the CTS and STS processes can achieve adjustable H_2:CO ratio with optimized value of 2:1 with little CO_2, unconverted hydrocarbons, and steam, which is required for down-stream processes to produce liquid fuels and chemicals.

559995 Subcritical and Supercritical Water Gasification of Soybean Straw and Flax Straw for Hydrogen Production: Experimental and Thermodynamic Modeling

Over the past few years there has been a tremendous demand for hydrogen (H_2) production for use as transportation fuels, desulfurization, heavy oil upgrading and in fuel cells. H_2 produced from waste biomass is widely considered as a green fuel useful for combined heat and power generation due to its high-energy content. Recently, lignocellulosic biomass has received immense attention as a promising feedstock for biofuel production due to its abundant availability, low cost and the ability to reduce atmospheric CO_2 through photosynthesis. Supercritical water gasification of lignocellulosic biomass is an attractive option for hydrogen production. The process is suitable for biomass with high moisture content because the medium of gasification is water. In addition, the syngas is produced at high pressure, thereby eliminating the need for gas compression. In this study, Soybean straw and Flax straw were characterized and examined as the feedstocks for H_2 production in subcritical (300^0C) and supercritical water (400^0C and 500^0C). To optimize the non-catalytic process, we studied the influence of temperature ($300 - 500^0$C), biomass to water ratio, BTW (1:5 and 1:10), biomass particle size (0.03 – 2 mm) and reaction time (30 – 60 min) on H_2 production at a pressure range of 22-25 MPa. Maximum H_2 yield and total gas yields of 6.62 mmol/g and 14.91 mmol/g, respectively were obtained at highest test temperature (500^0C), lower feed concentration (1:10 BTW), smallest particle size biomass (< 2 mm) and longest reaction time (45 min). A comparative evaluation of the gasification performance of the two lignocellulosic biomasses reveal that soybean straw exhibited superior H_2 yield (6.62 mmol/g) and total gas yield (14.91 mmol/g). Similarly, the gaseous products from soybean straw showed improved lower heating value (1592 kJ/Nm^3). To evaluate the drift of our experimental H_2 yield from the theoretical values, thermodynamic modelling using Gibbs free minimization method (GFMM) was performed. Since supercritical water gasification involves a series of complex reactions with several intermediates formation, the GFMM method was selected as it does not require the identification of intermediate reactions. The experimental results showed good correlation with the thermodynamic models, which will be discussed during the presentation. Our findings suggest that supercritical water gasification of lignocellulosic biomasses could be a useful green technology for effective H_2 production and waste valorization.

571809 A Compact System for Mass Production of Hydrogen from Biogas

Over the past decades, most of the energy has been produced from processes based on fossil fuels, and anthropogenic greenhouse gases generated as by-product have lead serious environmental problems. To address these issue, the alternative energy system based on the hydrogen so called as "hydrogen economy" has been highlighted. Conventional processes for the mass production of hydrogen are based on the catalytic gasification of biomass owing to maturity of the technology and favorable economic. Recently, the dry reforming of biogas (CH_4+CO_2) is considered as non-fossil fuel based hydrogen production reaction as well as carbon removal process consuming two major greenhouse components, i.e. CO_2 and CH_4. The effluent gas from the dry reforming reaction could be further converted to hydrogen via water gas shift reaction. In this study, a compact system that combines methane reforming and water gas shift reactions was experimentally demonstrated. From the suggested system, bulk hydrogen could be directly produced from biogas. Moreover, it is possible to not only produce high-purity hydrogen from a single system but also recycle captured CO_2 using solid sorbents in the reforming reaction to further improve the efficiency of the process by applying the sorption-enhanced reaction concept.

577341 A Dynamic Genome-Scale Metabolic Network Model for a Novel Methanotroph-Cyanobacteria Coculture

A Dynamic Genome-scale Metabolic Network Model for a Novel Methanotroph-Cyanobacteria Coculture

Kiumars Badr, Q. Peter He and Jin Wang*

Department of Chemical Engineering, Auburn University, Auburn, AL, 36849, USA. *Email: wang@auburn.edu

Biogas is comprised primarily of methane (50%~70%) and carbon dioxide (30% ~50%), which can be produced from various waste sources, including landfill material, animal manure; wastewater; and industrial, institutional, and commercial organic wastes. EPA estimates that currently US biogas production potential is 654 billion cubic feet per year, which could displace 7.5 billiongallon of gasoline [1]. It is clear that biogas has immense potential as a renewable feedstock for producing high-density fuels and commodity chemicals. However, the utilization of biogas represents a significant challenge due to its low pressure and presence of contaminants such as H_2S, ammonia, and volatile organic carbon compounds. To tap into this immense potential, effective biotechnologies that can co-utilize both CO_2 and CH_4 are needed.

In our previous work, we have demonstrated that metabolic coupling of methane oxidation to oxygenic photosynthesis can be a highly efficient way to recover the energy and capture carbon from biogas [1, 2]. However, development of multi-organism platforms for commercial biogas conversion present significant challenges which center around our ability to control function and composition of species in the Coculture. An essential tool for the optimization, design and analysis of the coculture based biogas conversion is the development and validation of kinetics models that can accurately describe and predict the co-culture growth under different conditions [3]. To this end, using *Methylomicrobium buryatense - Arthrosipira platensis* as the model coculture system, we have developed an unstructured model to capture the growth dynamics. Specifically, Monod-like models were developed to capture coculture growth. Two sources of substrate were considered in the model: gas transferred from gas phase and gas produced *in situ*. In addition, we rely on the fitted maximum cell growth rate for both strains to capture other potential interactions. Using designed experiments and the developed model, we clearly demonstrated that the synergistic effect within the coculture cannot be fully explained by the *in situ* substrate exchange, and there must be other "metabolic links" to explain the significantly enhanced cell growth of both strains in the coculture.

In order to test different hypothesized "metabolic links" between the Methanotroph and cyanobacteria in the Coculture, in the work, we developed a structured, dynamic genome-scale metabolic network model for the Coculture system. These GEMs, especially the dynamic GEMs, offer a comprehensive picture of cellular metabolism and serve as a bridge that can better link the work of microbiologists and engineers in understanding and optimizing complex cellular metabolism. By integrating the available knowledge on each strain with data obtained in our own experiments, we use DFBAlab to implement the dynamic GEM for the Coculture [4]. For the dynamic GEM, besides the GEMs for each individual strain, the key inputs to the DFBA model are the uptake kinetics for different substrates. In this work, the substrate update rates are provided from the unstructured dynamic model we already developed. In addition, the product secretion rates and cell growth rates predicted by the validated unstructured model were used to validate the coculture dynamic GEM. Although the experimental measurements to validate the model are mainly limited to cell growth, substrate consumption and product secretion rates for each organism in the coculture, the overall dynamic trajectory (both measured low-frequency samples and unstructured model predicted high-frequency samples) offers significantly more power in validating the model, compared to a few steady state conditions. The validated coculture GEM enables the testing of different hypotheses on potential "metabolic links" within the coculture.

Reference:

[1] Badr, K., Hilliard, M., He, Q.P., & Wang, J., "Understanding the Stability and Robustness of a Methanotroph-Cyanobacterium Coculture through Kinetic Modeling and Experimental Validation." Annual AIChE Meeting. Pittsburgh, PA (2018).

[2] Badr K, Roberts N, Hilliard M, He QP, Wang J. "Photoautotroph-Methanotroph Coculture – A Flexible Platform for Efficient Biological CO_2 – CH_4 Co-utilization," Dynamics and Control of Process Systems, including Biosystems - 12th DYCOPS. (2019); accepted.

[3] Almquist J., M. Cvijovic, V. Hatzimanikatis, J. Nielsen, M. Jirstrand, "Kinetic models in industrial biotechnology – improving cell factory performance", Metab. Eng., 24 (2014), pp. 38-60.

[4] Gomez JA, Höffner K, Barton PI. DFBAlab: a fast and reliable MATLAB code for dynamic flux balance analysis. BMC Bioinformatics. (2014),15:409.

567780 Kinetics and Reactor Design for Pyrolytic Treatment of Petroleum-Contaminated Soils

Scope and Significance

While marine oil spills from offshore platforms or tankers attract most of the public attention, the vast majority of oil spills (about 98%) occur on land. Every year about 10-25 million gallons of petroleum products are spilled mostly from pipelines and fixed facilities. Petroleum hydrocarbons pose long term threats to groundwater quality, inhibit the germination and root elongation rates of plants and are extremely toxic to important soil microbes. Moreover, polycyclic aromatic hydrocarbons (PAH) present in petroleum can damage the immune systems of aquatic organisms and wildlife.

Current remediation methods for petroleum-contaminated soils are either relatively slow or have unintended consequences in the form of soil damage and high-energy usage. Even worse, some processes such as aerobic bioremediation can activate toxic hydrocarbons and transform them to more noxious byproducts such as PAH derivatives. For all these reasons, there is still a pressing need for more efficient and *sustainable remediation* of petroleum-contaminated soils.

Pyrolysis is receiving increasing attention as an on-site remediation approach because of its potential to rapidly and reliably remove total petroleum hydrocarbons (TPH) with lower energy requirements and better post-treatment soil fertility than other *ex situ* thermal remediation approaches [1-3]. Pyrolytic treatment of contaminated soils at 420°C with only 15-min residence time in a continuous reactor reduced TPH by 99.9%, lowered the total PAH concentration by almost 95%, and restored soil fertility to 98% of the clean soil level [3]. These results clearly suggest that pyrolysis has the potential for improved ecosystem restoration following remediation compared to traditional thermal technologies. However, the same study also revealed potential tradeoffs between pyrolytic treatment intensity, soil detoxification efficacy and soil fertility restoration. Specifically, treatment at 470°C for 15- or 30-min reduced soil fertility to 51% and 39% of the clean soil level, which was only marginally higher that the fertility of the contaminated soil.

The existence of an optimal treatment intensity for contaminated soils underscores the need for mechanistic insight into the pyrolysis process and for a deeper understanding of how the detoxification of contaminated soil and the restoration of soil fertility are affected by (a) the properties of both the soil and the petroleum crude, and (b) the operating conditions of pyrolysis reactors.

Kinetics and Reactor Design for Pyrolytic Treatment of Petroleum-Contaminated Soils

Taking the first steps towards a rational design of pyrolysis reactors, we have combined mathematical techniques and advanced analytical measurements to:

1. Develop a full kinetic model for the pyrolytic treatment of soils contaminated with petroleum hydrocarbons. The new model describes all thermally-induced transformations of soil components (e.g. clays, carbonates, or SOM), the evaporation of light hydrocarbons and the pyrolysis of the heavier hydrocarbon fractions.
2. Develop a mathematical model describing the operation of continuous reactors (rotary kilns) that will be used for *ex situ* treatment of petroleum-contaminated soils. This dynamic model will then be used to estimate the optimal (a) pyrolysis temperature, (b) reactor residence time, and (c) gas flow rates that will achieve a specified TPH reduction at the minimum energy cost for any given throughput of contaminated soil.

Two clean soils with different compositions were spiked with oil crudes to prepare petroleum-contaminated soils. Thermogravimetry with evolved gas analysis (TG-IR and TG-MS) was used first to identify the mineral transformations taking place as the clean soils were heated in an anoxic atmosphere. The same analytical techniques were applied to study the two distinct processes occurring during the pyrolytic remediation of contaminated soils. Desorption of the lighter hydrocarbons is the dominant process for temperatures between 150 and 350°C. Thermal cracking reactions dominate in the 350-500°C range releasing pyrolysis products (like hydrogen, methane, or olefins) and leading to the formation of a carbonaceous solid material (char) that remains in the treated soil [2]. It is important to note that hydrocarbon desorption and pyrolysis overlap with the temperature ranges over which soil mineral transformations take place.

To develop our kinetic model, we assumed that the soil consists not only of inert components but also of "reacting pseudo-components" (e.g. clays, carbonates, SOM) that release either water or carbon dioxide as the soil is heated. Weight losses due to water or carbon dioxide release accounts for more than 95% of the weight loss observed during heating of clean soils. Contaminated soils contained additional pseudo-components that account for the weight losses observed due to the desorption of light hydrocarbons and the pyrolysis of the heavy petroleum fractions to yield gaseous products and char (coke). Using the distributed activation energy approach (DAE) approach, each pseudo-component was assumed to consist of a large number of species that react with activation energies distributed according to Gaussian probability density functions (PDF). To obtain the kinetic parameters, an optimization problem was formulated by fitting model predictions to experimental thermogravimetric data. Solving this nonlinear constrained optimization problem with an interior-point algorithm, the expected activation energies and variances of all pseudo-components were estimated, together with the preexponential factors for all the reactions and the initial mass fractions of all pseudo-components.

Excellent fits to the experimental data were obtained using four reacting pseudo-components for the clean soils. Three pseudo-components were required to model clay dehydration and dehydroxylation processes that occur over a wide range of temperatures and generate different multi-peak water release patterns for each soil. One pseudo-component was enough to model carbon dioxide release from either carbonate decomposition reactions or soil organic matter (SOM) pyrolysis. Two additional pseudo-components were required to model hydrocarbon desorption and pyrolysis.

The kinetic models were then used to simulate the operation of non-isothermal rotary kiln reactors that will be used for *ex situ* treatment of petroleum-contaminated soils. To preserve anoxic conditions, the reactors were heated externally while a stream of nitrogen gas flowed through the rotary kiln to sweep the desorbing hydrocarbons and pyrolysis products. A finite stage or mixing cell model with separate solid and gas phases was used to describe the operation of the pyrolysis reactor.

Model predictions were compared to experimental data we obtained with a continuous rotary kiln reactor [3]. The TPH reductions predicted by the model agreed very well with the experimental data for all the conditions considered in our earlier study: pyrolysis temperatures, 370, 420 and 470°C; residence times 15, 30 and 60 min; solid flow rates 7, 14, and 25 lb/h.

More importantly, however, our model was able to explain the trade-offs observed in our earlier study between pyrolysis treatment intensity, soil detoxification and soil fertility restoration [3]. The clay components of the soil used in [3] undergo significant dehydration between 350 and 450°C, a range that overlaps with that of pyrolysis reactions. As contaminated soil moves through the reactor zone kept at 420°C, the reaction rate for the heavy hydrocarbon fraction is fast enough to achieve almost complete conversion within 15 or 30 min. But, the same conditions are not severe enough to completely remove the water from the clay components of the soil. If the soil is treated at 470°C, on the other hand, the clay components are almost completely dehydrated, causing irreversible soil damage and loss of fertility.

By emphasizing the importance of soil composition and by raising the possibility of trade-offs between detoxification and fertility of treated soils, this study has significantly advanced our fundamental understanding of the *multiple* and *concurrent*physicochemical processes taking place during the pyrolytic treatment of petroleum-contaminated soils. Theoretical kinetic and reactor models are essential for designing pyrolysis reactors optimized for specific petroleum-soil systems, enabling robust operation and facilitating acceptance of this novel technology by regulatory agencies and stakeholders.

References

1. Vidonish, J. E.; Zygourakis, K.; Masiello, C. A.; Gao, X.; Mathieu, J.; Alvarez, P. J., *Environmental Science & Technology* **2016,** *50*, (5), 2498-2506.
2. Vidonish, J. E.; Alvarez, P. J. J.; Zygourakis, K., *Industrial & Engineering Chemistry Research* **2018,** *57*, (10), 3489-3500.
3. Song, W.; Vidonish, J. E.; Kamath, R.; Yu, P. F.; Chu, C.; Moorthy, B.; Gao, B. Y.; Zygourakis, K.; Alvarez, P. J. J., *Environmental Science & Technology* **2019,** *53*, (4), 2045-2053.

573966 Well Placement Optimisation in a Heterogeneous Petroleum Reservoir with Geological Uncertainty

ABSTRACT

Huge investments are required in the development of oil and gas fields; this implies that decisions at any stage of the development process must be aided by sound and proven mathematical-oriented methods, because the use of intuitive engineering judgement alone cannot guarantee sustainable profitability over long periods [1]. This presents an ever-increasing motivation to apply optimisation techniques for the enhancement of petroleum production from both planning and operational/scheduling perspectives. The planning aspect typically involves activities such determining the number and locations of wells, drilling the wells and infrastructure installation. On the other hand, the operational/scheduling aspect comprises of well rate allocation, flow scheduling and routing problems [2]. Incorporating optimisation at the planning stage is a pre-emptive step that mitigates subsequent operational difficulties over the field's lifetime. However, embedding optimisation in a planification problem can be more challenging compared to production and injection scheduling optimisation in an oil and gas field. Although many studies [3, 4] have addressed the operational/scheduling aspect of the field development problem, this work is premised on the proposition that, jointly optimising the planning and scheduling problem better enhances field operability.

The uncertainty of subsurface geology is another constraint to take into consideration when optimising field productivity [5]. To capture the uncertainties in the subsurface geological (reservoir) model, geostatistical realisations of the most accurate reservoir representation are obtained using available information (permeabilities, porosities, fluid saturation) [6]. These uncertainties in the model are often reflected in the diverse outcomes of well productivities and ultimately the field's Net Present Value (the objective function) [7, 8]. Given the numerous number of possible realisations, it is impractical to consider all realisations in the optimisation; hence, it is important to apply classical reduction methods to ensure computational feasibility.

In this work, we specifically address the well placement planning and rate control optimisation problem under geological uncertainty. The difficulty of this problem is characterised by the presence of discrete variables, nonlinear and nonconvex objective function and constraints. Increased complexity arises from the computationally demanding functional evaluations from full-scale reservoir simulation, which is done in ECLIPSE®. We also exploit the MATLAB® Reservoir Simulation Toolbox – MRST® [9] for interfacing with ECLIPSE® and performing derivative-free optimisation tasks. The developed workflow is applied to a synthetic case study, for which it is demonstrated using several operational scenarios that using optimally selected realisations reduces computational time while providing best operating well locations. A comparative analysis of the performance of different derivative-free optimisation algorithms is also presented.

REFERENCES

1. Epelle, E.I. and Gerogiorgis, D.I., Optimal Rate Allocation for Production and Injection Wells in an Oil and Gas Field for Enhanced Profitability. *AIChE J.,* **in press** (DOI: 10.1002/aic.16592).
2. Tavallali, M.S. et al., 2014. Well placement, infrastructure design, facility allocation, and production planning in multireservoir oil fields with surface facility networks. *Ind. Eng. Chem. Res.*, 53(27), 11033-11049.
3. Gunnerud, V., Conn, A. and Foss, B., 2013. Embedding structural information in simulation-based optimization. Comput. *Chem. Eng.,* 53, 35-43.
4. Kosmidis, V.D., Perkins, J.D. and Pistikopoulos, E.N., 2004. Optimization of well oil rate allocations in petroleum fields. *Ind. Eng. Chem. Res.,* 43(14), 3513-3527
5. Li, L., Jafarpour, B. and Mohammad-Khaninezhad, M.R., 2013. A simultaneous perturbation stochastic approximation algorithm for coupled well placement and control optimization under geologic uncertainty. *Comput. Geosci.*, 17(1), 167-188.

6. Wang, H., Echeverría-Ciaurri, D., Durlofsky, L. and Cominelli, A., 2012. Optimal well placement under uncertainty using a retrospective optimization framework. *SPE J.*, 17(01), 112-121.
7. Bellout, M.C. et al., 2012. Joint optimization of oil well placement and controls. *Comput. Geosci.*, 16(4), 1061-1079.
8. Gerogiorgis, D.I., Georgiadis, M., Bowen, G., Pantelides, C.C. and Pistikopoulos, E.N., 2006. Dynamic oil and gas production optimization via explicit reservoir simulation. *Comput. Aided Chem. Eng.*21, 179-184.
9. Lie, K.A., 2014. An introduction to reservoir simulation using MATLAB: user guide for the Matlab Reservoir Simulation Toolbox (MRST). *Sintef ICT*, Trondheim, Norway.

569460 Modeling of Thermal Cracking of Natural Gas Liquids: Optimization with Support Vector Machine-Based Constraint Handling Scheme for Stiff ODEs

Rapid increase in shale gas production in the Appalachian Basin (Marcellus and Utica Shales) for the past 5 years has led to a significant growth in natural gas liquids (NGLs) production as well as a projected increase of these petrochemical feedstocks in the upcoming years [1, 2]. In this perspective, many existing ethylene crackers have expanded capacity and new crackers are becoming online to benefit from this unique opportunity [3]. Hence, the mathematical modeling and optimization of this process emerges as a necessity so as to determine the optimal operating conditions and the reactor configurations for the steam cracker, in such a way that the profit from thermal cracking of NGLs is maximized.

In this work, we are demonstrating (i) the comprehensive first-principles modeling of an adiabatic propane steam cracking process as a 1D plug flow reactor with coking effects [4] and (ii) the data-driven optimization of this reactor model using the ARGONAUT [5-6] framework. The mathematical model for thermal cracking includes the reaction rates expressed with Arrhenius equation, as well as the ODEs for continuity, energy, and momentum balances. The kinetics of thermal cracking of propane is adapted from Sundaram and Froment (1979) and the cracking model is validated using industrial data [7]. This simulation model is later used in the data-driven optimization phase to collect input-output data. However, the stiffness in the reactor models poses an additional complexity in retrieving the optimal operating conditions for the cracking process, especially when the data-driven optimization strategies strongly rely on collecting numerically stable samples from the reactor simulator. To this end, a novel Support Vector Machine (SVM)-based filtering approach is applied to eliminate candidate sampling points that are susceptible to the stiffness in the model and that will result with an abrupt termination of the cracking simulation due to high pressure drop. The SVM classifier identifies and removes the numerically unstable candidate points *a priori* to simulation execution by formulating an optimal separating hyperplane between stable/unstable points. This enables the data-driven algorithm to consistently converge to a profitable and feasible reactor configuration for the propane cracking process with an improved objective function value.

References

[1] Energy Information Administration. Today in Energy: Appalachia, Permian, Haynesville drive U.S. natural gas production growth, August 2018. URL https://www.eia.gov/todayinenergy/detail.php?id=36934. (accessed 04.10.2019).

[2] Energy Information Administration. Short-term energy outlook, April 2019. URL https://www.eia.gov/outlooks/steo/. (accessed 04.10.2019).

[3] Energy Information Administration. U.S. ethane consumption, exports to increase as new petrochemical plants come online, February 2018. URL https://www.eia.gov/todayinenergy/detail.php?id=35012. (accessed 02.26.2018).

[4] O. Onel. (2017). Advances in Modeling, Synthesis, And Global Optimization of Hybrid Energy Systems Toward the Production of Liquid Fuels and Olefins. Princeton University.

[5] F. Boukouvala and C. A. Floudas. (2017). ARGONAUT: AlgoRithms for Global Optimization of coNstrAined grey-box compUTational problems. *Optimization Letters*, 11(5), 895-913.

[6] B. Beykal, F. Boukouvala, C. A. Floudas, N. Sorek, H. Zalavadia, E. Gildin. (2018). Global optimization of grey-box computational systems using surrogate functions and application to highly constrained oil-field operations. *Computers & Chemical Engineering*, 114, 99-110.

[7] K. M. Sundaram and G. F. Froment. (1979). Kinetics of coke deposition in the thermal cracking of propane. *Chemical Engineering Science*, 34, 635-644.

574887 Towards a Novel Energy Price Predictive Framework: The Texas A & M Energy Price Index

The energy markets are sensitive and volatile to technological breakthroughs, changes in monetary policy, or major global challenges. Moreover, various governmental agencies, political and commercial organizations, think tanks as well as researchers and academics around the globe, consider various energy policies and their effects in an attempt to deal with the increasing concerns in energy independence, energy scarcity, energy sustainability, and pollution caused by the utilization of energy. Also, in the current environment in which all strategic political and commercial decisions and policies are assessed in economic terms, it is of outmost importance to determine the cost of energy precisely so as to evaluate their effectiveness. Considering the fact that energy affects every single individual and entity in the world, it is crucial to quantify "the unit price of energy" accurately, and study how it evolves with respect to time, major breakthroughs, challenges, energy and monetary policies. Given the lack of such a tool, we develop a novel framework, the Texas A&M Energy Price Index (EPI) that can be used as a benchmark to calculate the average price of energy to the end-use consumers in the United States.

The complicated energy landscape of the United States is carefully depicted to determine the products that are directed to the end-use sectors of the U.S. economy [1-5]. The total energy demand of these products, together with their prices, serve as the backbone of the Energy Price Index [6-10]. Since several key components (e.g. prices, demands) of the Energy Price Index have a lag between two and three months, we introduce a rolling horizon based parameter estimation model to approximate the current value of the Energy Price Index. The predictive methodology is tested on 120 months to show the accuracy of the developed framework in the prediction of the current value of the Energy Price Index. Energy price sub-indices for the transportation, residential, commercial, and industrial sectors are also developed. Some of the potential applications of the proposed index in the areas of economics, finance, engineering, law, and policy are demonstrated.

References:

- Elia, J. A.; Baliban, R. C.; Xiao, X.; Floudas, C. A. Optimal energy supply network determination and life cycle analysis for hybrid coal, biomass and natural gas to liquid (CBGTL) plants using carbon-based hydrogen production. Computers and Chemical Engineering 2011, 35, 1399-1430.
- Elia, J. A.; Baliban, R. C.; Floudas, C. A. Nationwide energy supply chain analysis for hybrid feedstock processes with significant CO2 emissions reduction. AIChE Journal 2012, 58 (7), 2142-2154.
- Elia, J. A.; Floudas, C. A. Energy Supply Chain Optimization of Hybrid Feedstock Processes: A Review. Annual Review of Chemical and Biomolecular Engineering 2014, 5 (1), 147-179.
- Onel, O.; Niziolek, A. M.; Floudas, C. A. Integrated biomass and fossil fuel systems towards the production of fuels and chemicals: state of the art approaches and future challenges. Current Opinion in Chemical Engineering 2015, 9, 66-74.
- Floudas, C.A.; Niziolek, A. M.; Onel, O.; Matthews, L.R. Multi-Scale Systems Engineering for Energy and the Environment: Challenges and Opportunities. AIChE Journal 2016, 62 (3), 602-623.
- Energy Information Administration (EIA), Annual Energy Outlook, 2019
- Energy Information Administration (EIA), Monthly Energy Review, 2019
- Lazard, Levelized Cost of Energy Analysis Version 12.0, 2018
- Bureau of Labor Statistics (BLS), Producers Price Index, 2018

U.S. Department of Energy (DOE), Alternative Fuels Data Center, 2018

575666 Intensification of Biorefinery Operations Using Building Blocks

Bioenergy is an attractive form of energy since it requires renewable sources that can be produced locally and used as raw materials for its production. One of the possible routes for bioenergy production is through lignocellulosic biorefineries, which utilize various biomass sources, e.g. corn stover, wood, rice straw and forest wastes, to produce biodiesel, steam and power in addition to pulp and paper [1]. Increasing awareness towards sustainable and modular technologies highlights the need for novel processing pathways leveraging on process intensification principles that can yield drastic improvements in several metrics, including economics, energy consumption and chemical footprint [2]. Recent developments in process intensification resulted in development of Integrated biorefinery concept with two new processing pathways: (i) pre-extraction of hemicellulose from biomass and its conversion to ethanol, and (ii) production of syngas from gasification of black liquor [3]. However, there exist further unexplored opportunities for intensification of the biorefineries to enhance the reaction yield and mass/heat transfer within the several parts of the process, e.g. digestion of pulp, evaporation of black liquor, etc. [4].

Recently, a new method for systematic process intensification, which relies on building block-based representation of chemical process flowsheets, has been introduced [5]. These building blocks can be integrated to form a two-dimensional grid embedding numerous non-trivial process alternatives [6-9]. Each block can be assigned with several physicochemical phenomena, e.g. reaction, vapor-Liquid equilibrium, solid-liquid equilibrium, etc., combination of which can yield various intensification pathways. In this work, we extend this novel representation methodology to various operations that take place in a biorefinery. Pulping which occurs in a pulp digester lies at the heart of a pulp mill and, here, we develop a new representation based on fundamental building blocks to represent the pulp digestion process. Mass transfer boundary layer is represented by the boundary shared by two neighboring blocks in different phases and digestion process is represented within the interior of the block. This building-block based representation is modeled using a mixed-integer nonlinear programming (MINLP) problem so as to minimize the energy consumption and increasing the yield of the process while searching for several intensified pathways. The proposed approach is demonstrated through several case studies focusing on the conventional Kraft pulping process, Kraft pulping process with pre-extraction of hemicellulose, Kraft pulping process with pre-extraction of hemicellulose and short fibers.

[1] Huang, H., Lin, W., Ramaswamy, S., Tschirner, U. (2009) Process Modeling of Comprehensive Integrated Forest Biorefinery – An Integrated Approach. *Applied Biochemistry and Biotechnology, 154*, pp. 205-216.

[2] Tian, Yuhe., Demirel, S.E.., Hasan, M.M.F., Pistikopoulos, E.N., (2018) An overview of process systems engineering approaches for process intensification: State of the art. *Chemical Engineering and Processing - Process Intensification, 133*, pp. 160-210.

[3] Aden, A., Ruth, M., Ibsen, K., Jechura, J., Neeves, K., Sheehan, J., Wallace, B., Montague, L., Slayton, A., Lukas, J., (2002) NREL Report TP-510-32438. Lignocellulosic Biomass to Ethanol Process Design and Economics Utilizing Co-Current Dilute Acid Prehydrolysis and Enzymatic Hydrolysis for Corn Stover.

[4] Huang, H., Ramaswamy, S., Al-Dajani, W.W., Tschirner, U. (2010) Process modeling and analysis of pulp-mill based integrated biorefinery with hemicelluloses pre-extraction for ethanol production: A comparative study. *Bioresource Technology 101*, pp. 624-631.

[5] Demirel, S.E., Li, J., Hasan, M.M.F., (2017) Systematic process intensification using building blocks. *Computers and Chemical Engineering, 105*, pp. 2-38.

[6] Li, J., Demirel, S. E., Hasan, M. M. F., (2018) Process Synthesis Using Block Superstructure with Automated Flowsheet Generation and Optimization. *AIChE Journal, 64(8)*, pp. 3082-3100.

[7] Li J., Demirel S.E., Hasan M.M.F. (2018) Process Integration using Block Superstructure. *Industrial &*

Engineering Chemistry Research, *57(12)*, pp. 4377–4398.

[8] Li, J., Demirel, S.E., & Hasan, M. M. F. (2019). Building Block-Based Synthesis and Intensification of Work-Heat Exchanger Networks (WHENS). *Processes, 7(1),* 23.

[9] Demirel, S. E., Li, J., and Hasan, M. M. F. (2019). A General Framework for Process Synthesis, Integration, and Intensification. *Industrial & Engineering Chemistry Research.* DOI: 10.1021/acs.iecr.8b05961.

576096 Hyperbolicity of Heat Transport Processes

Heat transport models are ubiquitously present in chemical, petrochemical, manufacturing and process industry as mathematical models of distributed parameter systems given by parabolic partial differential equations (PDEs). The conductive (diffusive) transport over macroscopic length scale is characterized by the well known Fourier's law and a large number of research efforts have been realized in the realm of modelling, control and optimization of such models. However, the salient characteristic of these models is that any initial disturbance in the material body is propagated instantly due to the parabolic nature of the partial differential equation (PDE) model. To eliminate this unphysical feature, we revisit the Fourier law to take into account the thermal inertia, resulting in a hyperbolic heat equation which avoids the phenomenon of infinite speed of propagation of initial data [1,2].

The main conservation laws are embedded in a modelling variety provided by the hyperbolicity of the transport systems which is physically relevant and desired property as action at distance is precluded and physically meaningful finite speed of phenomena propagation is ensured. Furthermore, the hyperbolicity mathematically ensures the well-posedness of local Cauchy problems [3]. In addition, a Stefan problem is considered, which is a specific type of heat transport boundary value problem for a partial differential equation describing the heat distribution evolution in a phase changing medium [4]. Since the moving interface is unknown a priori, the solution needs to take into account determination of the moving boundary position as well as accurate heat propagation. Finally, we provide derivation of the hyperbolic heat transport equations, the stability analysis and the numerical results for simple heat diffusion problem on fixed domain and Stefan problem.

Literature Cited

1. Cattaneo C. On a form of heat equation which eliminates the paradox of instantaneous propagation. *C R Acad Sci Paris*. 1958;pp. 431–433.

2. Vernotte P. Les paradoxes de la theorie continue de l'equation de la chaleur. *C R Acad Sci Paris*. 1958;246:3154–3155.

3. Fischer AE, Marsden JE. The Einstein evolution equations as a first-order quasi-linear symmetric hyperbolic system, I. *Communications in Mathematical Physics*. 1972;28(1):1–38.

4. Gupta S. *The Classical Stefan Problem. Basic Concepts, Modelling and Analysis*. North-Holland: Applied mathematics and Mechanics. 2003.

569568 Constructing Efficient Surrogate Models for High-Dimensional First Principles Systems Under Uncertainty

A plethora of chemical engineering processes are informed by sophisticated numerical simulations based on first-principles models (e.g., consisting of a set of coupled partial differential equations) that exhibit highly nonlinear behavior and are computationally expensive. It is often the case that the parameters in these models as well as the initial and boundary conditions in dynamical systems are uncertain and, thus, this uncertainty is propagated to the model outputs. Therefore, relying on deterministic simulation results could possibly lead to erroneous outcomes for the design, operation, and control of these processes. The computational burden associated with solving the entire set of equations can be prohibitive when one wishes to run the model for the whole spectrum of possible input parameter values, even in this era of high-performance computing, especially in the context of optimization applications. Metamodeling techniques aim at alleviating this problem by approximating first-principles model responses under uncertainty with efficient black-box mappings that can serve as surrogate models for the original high-fidelity models [1]. The primary notion in metamodeling is to calibrate the surrogate models using the fewest possible runs of the original model without compromising their predictive capability.

In this work, we explore the effectiveness of various well-established meta-modeling techniques applied to a basic, but relevant crystallization problem, i.e., the growth/dissolution of a 2D crystal population through cooling/heating cycles. This technique is used, possibly in combination with milling, to manipulate the crystal shape of active pharmaceutical ingredients, and efforts have been recently devoted to optimize and control this process. Even though relatively simple, such a problem represents an interesting test-case for metamodeling since it is highly nonlinear, inherently dynamic, and industrially relevant. [2,3] The crystallization system is described by a system of integro-partial differential equations, whose parameters are inherently uncertain. The parameter uncertainties are modeled by suitable probability distributions selected according to available experimental data and expert knowledge.

We first employ sparse polynomial chaos expansions (PCE) whose coefficients are estimated based on least-angle regression and explore their scalability with an increasing number of uncertain inputs. PCEs have been traditionally used for uncertainty quantification in numerous applications (e.g., see [4]). We then present the fundamentals of Gaussian Processes (i.e., Kriging) [5] and explore their coupling with sparse PCEs in order to create surrogate models with optimal local and global predictive capabilities [6]. Finally, the use of NARX neural networks [7] is demonstrated to deal with the continuous dynamic behavior of the system. The accuracy of these three surrogate modeling approaches in capturing the highly nonlinear spatio-temporal dynamics of the crystallization process at hand is assessed using state-of-the-art cross-validation and validation measures, and their computational aspects are discussed.

[1] Bhosekar A., Ierapetritou M., “Advances in surrogate based modeling, feasibility analysis, and optimization: A review”, Computers and Chemical Engineering, 108:250-267, 2018

[2] Salvatori, F., Mazzotti, M., “Manipulation of particle morphology by crystallization, milling, and heating cycles—A mathematical modeling approach”, Industrial and Engineering Chemistry Research, 56:9188–9201, 2017.

[3] Bötschi, S., Rajagopalan, A. K., Morari, M., Mazzotti, M., “Feedback control for the size and shape evolution of needle-like crystals in suspension. I. Concepts and Simulation Studies”, Crystal Growth and Design, 18:4470–4483, 2018.

[4] Paulson J.A., Mesbah A., “An efficient method for stochastic optimal control with joint chance constraints for nonlinear systems”, International Journal of Robust and Nonlinear Control, 1-21, 2017.

[5] Kawai S., Shimoyama K. “Kriging-model-based uncertainty quantification in computational fluid dynamics. 32nd AIAA Applied Aerodynamics Conference, 2014, 2737.

[6] Schöbi R., Sudret, B., Wiart J., “Polynomial-Chaos-based Kriging”, International Journal for

Uncertainty Quantification, Begell House Publishers, 5:171-193, 2015.

[7] Mai C.V., Spiridonakos M.D., Chatzi E.N., Sudret B., "Surrogate modeling for stochastic dynamical systems by combining nonlinear autoregressive with exogenous input models and polynomial chaos expansions", International Journal for Uncertainty Quantification, 6:313-339, 2016.

AUTHOR INDEX

AUTHOR INDEX

AUTHOR INDEX

AUTHOR INDEX

AUTHOR INDEX

AIChE
120 Wall Street, FL 23
New York, NY 10005-4020

CURRAN ASSOCIATES INC.
proceedings.com

ISBN 978-1-7138-0562-5